Claudia Verónica Reyes Guzmán
Leonor Muñoz Ramirez
Yinady Yarime Castillo Lazarin

Sodium Cyanide Adsorption on Different Materials

Claudia Verónica Reyes Guzmán
Leonor Muñoz Ramirez
Yinady Yarime Castillo Lazarin

Sodium Cyanide Adsorption on Different Materials

Molybdenite and coconut shell carbon in cyanide adsorption

Imprint

Any brand names and product names mentioned in this book are subject to trademark, brand or patent protection and are trademarks or registered trademarks of their respective holders. The use of brand names, product names, common names, trade names, product descriptions etc. even without a particular marking in this work is in no way to be construed to mean that such names may be regarded as unrestricted in respect of trademark and brand protection legislation and could thus be used by anyone.

Cover image: www.ingimage.com

This book is a translation from the original published under ISBN 978-620-3-88255-1.

Publisher:
Sciencia Scripts
is a trademark of
Dodo Books Indian Ocean Ltd. and OmniScriptum S.R.L Publishing group
Str. Armeneasca 28/1, office 1, Chisinau MD-2012, Republic of Moldova, Europe
Printed at: see last page
ISBN: 978-620-5-37094-0

Contents

To elaborate vegetable carbon from coconut husk, with this we intend to take it to the carbonisation stage transforming it into activated carbon in such a way that when it is in contact with aqueous cyanide solutions, it behaves like a mineral carbon and adsorbs the free cyanide.

Specific Objectives

1 Production of activated carbon from coconut husk residues using the Retsch mill.

2 Physical and chemical treatment of coconut husk residues from mechanical milling.

3 Adsorption of cyanide in the presence of modified and unmodified coconut shell activated carbon and comparison with other materials not normally used in adsorption.

4 Characterisation by X-ray Photoelectron Spectroscopy and Fourier Transform Infrared Spectroscopy of the best adsorbent.

INTRODUCTION

CN- is one of the toxic chemicals found in wastewater discharges from electroplating, metal finishing, steel hardening, mining (extraction of metals such as gold and silver), automotive parts manufacturing, photographic, pharmaceutical and carbon processing units, food and chemical synthesis (nylon, fibres, resins, fertilisers, pesticides and herbicides). Cyanide release from industries has been estimated to be over 14 million kg/year worldwide and can exist in three forms: total cyanide (CNT), cyanide WAD (Weak acid dissociable) and free cyanide (CNL).

Cyanide, when emitted in concentrations that exceed environmental regulations, is a very dangerous compound for humans, aquatic organisms and the environment in general, because it exerts an inhibitory action on certain metabolic enzymes; the main compound affected is the enzyme cytochrome c oxidase; its inhibition causes the blocking of the electron transport chain at the mitochondrial level, preventing the absorption of oxygen. According to Azamat and Khataee (2017), between 0.5 and 3.5 mg of cyanide per kilogram of body mass in humans can cause death.

The United States Environmental Protection Agency (USEPA) has proposed a regulation for drinking water and aquatic waters with respect to CNT of 0.2 and 0.05 mg/L, respectively, and in terms of CNL for the protection of aquatic life in fresh water, it establishes it at 0.022 and 0.0052 mg/L. The Peruvian regulations for water quality standards (D. S. N° 002 -2008 - MINAM), establishes the CN WAD at 0,1 mg/L (category 3: irrigation of vegetables and animal drinks) and for CNL at 0,022 mg/L (category 4: conservation of the aquatic environment).

Cyanide-containing chemical emissions above the permissible limits are treated by conventional chemical methods such as: alkaline chlorination, INCO (SO2/air) process (International Nickel Company's), hydrogen peroxide, Caro's acid method (H2SO4/H2O2), ozonisation, reverse osmosis, activated carbon, resins, among others; however chemical processes to degrade cyanide are not appropriate for environmental and economic perspectives; because of their high demand for chemicals and the formation of secondary pollutants, which need additional treatment before disposal and argues that adsorption is a simple and attractive method for the removal of toxic compounds from effluents due to its high efficiency, easy handling and economic feasibility.

According to literature references, several adsorbents have been studied; with and without treatment to adsorb CN- namely: olive pit and coffee residues, shea kernel husk, coconut husk, rice husk, almond husk, bituminous activated carbon and in the present study, minerals (molybdenite, calcite, fluorspar, barite) from a flotation exhauster and coconut husk carbon treated by a little studied method in the carbonisation and activation of the precursor were used.

Thus, the purpose of the present work was to evaluate the adsorption process of cyanide, contained in aqueous solution and put in contact with these materials.

2.1- GENERAL CHARACTERISTICS OF BARITE ORE

The main deposits of BARITE are associated with dolomite or limestone. Its usefulness is based on its high specific weight (4.3 to 4.6 gr/cc) which is modified by impurities. It is mainly used in the drilling of wells, because of its higher weight it descends to the bottom of the borehole, and forces the lighter muds to rise to where they can be extracted. It finds application in the glass industry, as it imparts homogeneity to glass, and in the chemical industry for the manufacture of various chemical compounds which, in turn, are used in paints, rubbers and inks. It is mainly associated with hydrothermal processes, either as a vein filler or gangue or as cavity fillers, especially in calcareous rocks, with which it is often associated [1].

Barite is a barium sulphate (BaSO4) and is found in nature as crystalline masses in whitish, greenish, greyish or reddish colours. Celestite (SrSO4) has the same crystal shape and structure as baryte. Sometimes celestite is blue in colour, another way to identify it is that it is of a lower density than barite. Additionally, a flame test can distinguish them: by exposing powdered crystals to fire, the colour of the flame will confirm the identity of the crystal, if it is pale green it is baryte, if it is red it is celestite [2].

Natural barite powders are chemically inert, easily dispersed, have low abrasion and excellent resistance to heat and corrosion, have low oil absorption, and act as a texturising agent in paints without the risk of damaging their lustre. Precipitated barium sulphate consists of ultrafine particles, is of high purity and acts as a pigment dispersant in coloured systems, while increasing the productivity of these pigments.

It is an inexpensive, clean, relatively soft mineral and makes up approximately 40% of the constituents of drilling mud. It has a light colour and high lustre (90%). Low oil absorption and oil wetting capacity allow it to be used as a filler and heavy agent in acoustic components, adhesives and sporting goods.

The brightness of barite can be increased by bleaching with sulphuric acid. Other relevant properties include its thermal stability (1580°C), thermal conductivity (6X10-3 cal/cm), specific heat (0.11 cal/g°C), dielectric constant (7.3) and coefficient of thermal expansion ($10X10^{-6}$).

Barite is a source of barium oxide in the manufacture of glass in which it acts as a flux, oxidiser and decolouriser, giving the glass a shinier, clearer finish. Barite also absorbs gamma radiation and can replace the charge in nuclear shields.

Barite is a material that aids environmental protection and has many advantages, such as strong inertness, good stability, resists acids and alkalis, and moderates stiffness.

PHYSICAL CHARACTERISTICS:

Crystals commonly tabular, also globular, fibrous or lamellar, penachite; coarsely laminated, resembling white marble and earthy, sporadically banded colours as in stalagmite.

It has a perfect cruciform, irregular fracture, brittle. Its hardness varies from 2.5 to 3.5 on the Mohs scale; specific gravity from 4.3 to 4.6 gr/cc; lustre is glassy, resinous, sometimes pearly. Its streak is white, its colour is very varied: white, yellowish white, grey, blue, red or brown, dark brown; transparent to translucent, translucent to opaque. When rubbed it can emit a fetid odour. [3]

Barite is the most common of the barium-bearing minerals; it occurs sometimes in large quantities as veins or layers; also as a gangue mineral in various mineral veins; or in crystals and crystal clusters. It is generally associated with lead, copper, iron, zinc, silver, nickel, cobalt, manganese and other ores. It is also associated with fluorite, quartz, calcite, dolomite, siderite, among others. Due to its characteristics, it is associated with a considerable amount of other minerals.

QWMICAL CHARACTERISTICS:

Barium compounds show close relationships with compounds of the other alkaline earth metals: calcium and strontium. As the atomic weight increases from calcium to barium, the specific gravity of the metal and the solubilities of the hydroxides increase; while it decreases for the halides, nitrates and sulphates. The solubilities of the barium salts are Upic of the alkaline-earth group: they are fairly soluble in acetate, chloride, bromide, iodide and nitrate; and insoluble in carbonate, chromate, fluoride, oxalate, phosphate and sulphate. All salts become more soluble to a greater or lesser degree as the pH decreases. With the exception of barium sulphate, they dissolve partially in carbonaceous acid and completely in chlor^dic or nitric acid. Sulphate is highly insoluble and is used for the determination of the barium ion. The difference in solubility between sulphate and carbonate is the basis for the considerable use of the latter in the ceramics industry. The most commonly used barium compounds are sulphate (barite) and carbonate (witherite), because the whole barium metallurgy is based on them.

SPECIFICATIONS:

Drilling mud grade (heavy barite)

Density	4.2 g/cm3 min.
Ca	250 ppm max.
Residue>75p	3% by weight, % max.
Residue>45p	Not specified.
Particles>6p in spherical equivalent diameter	30% by weight, % max.

Paint Filler Grade

BaSO4	95% min.
Fe2O3	0.05% max.
Foreign material	2.0%
Humidity	0.5%
Water-soluble compounds	0.2%
Particle size	99.98% -37pm (400 mesh) or Hegman 6.5
Brightness	80% +
Oil absorption	5kg/45 kg.
pH	6.4

Chemical grade barite powder:

BaSO4 96% min.
SrSO4 0.7% max.

CaSO4 0.6% max.
SiO23.5% max.
Whiteness range: 88% mm.
Powder size: 325 mesh

Chemical grade barite lump:

BaSO4 96% min.
SrSO4 0.7% max.
CaSO4 0.6% max.
SiO2 3.5% max.
Whiteness range: 80% mm.

Drilling grade raw barite:

Sperm gravity: 4.23 mm.
Calcium: 250 ppm max.
Mercury (Hg): 1.0 ppm max.
Cadmium (Cd): 3.0 ppm max.
Carbonates: 2500 mg/l max.
Sulphides: 50 mg/l max.
Humidity: 1.0% max.
Piece size: (0-150mm): 95% mm.

Glass grade barite:

BaSO4   95% mm.
SiO2   1.5% max.
Fe2O3   0.15% max.
            0.15% max.
Al2O3
<850 pm   100.0%
<150 pm   5.0% max.

Barium carbonate:
(BaCO3)   92-98% min.
Fe2O3   1% max.
SrSO4   1% max.
CaF2   0.5% max.

Barium sulphate precipitate:

BaSO4   98.8
Fe2O3   0.004
Sulphur   0.003
Water-soluble   0.20
Acid soluble   0.80
Humidity   0.20
Whiteness   90
                15.25
Oil absorption %
pH range   6.5-8.0

2.1.1 COMMERCIAL VARIETIES

Heavy spatula

It is a mineral constituted by barium sulphate, of light colour, medium hardness and high weight. According to its phosphorescence, high density, intense and resinous lustre and insolubility in acids, it is considered for:

Consumer activities:
- Raw material in drilling muds.
- Glass industry.
- Filling of paints.
- Manufacture of ceramics and paper
- X-ray examination (gastrointestinal tract).
- Automotive applications.
- Raw material for barium compounds.

Barium carbonate

Barium carbonate ($BaCO_3$), also known as witherite, is a chemical compound used in the manufacture of bricks, ceramic glazes and cement. The difference in solubility between sulphate and carbonate, when heated with mineral carbon, is the basis for the considerable use of this compound:

Consumer activities:
- Glaze fluxes
- Ceramics
- Foundry
- Optical glass
- Fine glassware
- Electronic application (cathode tubes)

Natural barite powder

Natural barium sulphate, also known as barite, is a classic filler mineral used in the paint production. The paint industry mainly uses grades with a BaSO4 content of more than 90%.

Among its advantages are its easy dispersibility, low abrasion and excellent resistance against heat and corrosion. Due to its extremely low oil absorption values, it is predestined for paint formulations. In this industry, it acts as a texturising agent without the risk of damaging its lustre. It has advantages in the mixing of pigments, in which in many cases it is very bright, for which micronised barite is used.

Barite with selected particle size distribution can be used to obtain lustre control in coatings.

Barium sulphates precipitate/fixed white

Obtained by a chemical precipitation process, allowing special chemical purity and ultra-fine particle sizes. Fixed white enhances brilliance in many coatings and acts as a pigment dispersant in coloured systems between white and coloured particles and increases the efficiency of coloured pigments.

In medical X-ray tests, the fixed target serves as a contrast agent, as well as for the crystallisation of lead sulphates in accumulators to maintain charge capacity. Resistant and high lustre papers are generally produced with precipitated barium sulphate. Thermoplastics can be filled with precipitated barium sulphate to obtain special characteristics.

2.1.2 GEOLOGICAL MINING POTENTIAL

Figure 1.- Map of Mexico with the most important baryte production points.

Barite production in Mexico is located in three geological provinces located in the north of the country: Nuevo Leon, Sonora and Coahuila, together these three states produce more than 98% of the mineral. In 2017, the mining region with the highest barite production potential is Nuevo Leon (Galeana), with 82.5% of the national production, tripling its production volume compared to 2016; followed by Sonora (Villa Pesqueira), which participates with 11.1% of the production, which presented a significant increase of 41.3% compared to 2016. The third producing entity is Coahuila (Muzquiz and Parras), which participates with 5%, presenting a slight decrease in its production volume (-28%) compared to 2016. There is also a minimal production of the mineral in other states that, together, generate 1.4% of the barite extracted in the country. Globally, Mexico ranks seventh in the production of this mineral.

2.1.3 barite production process

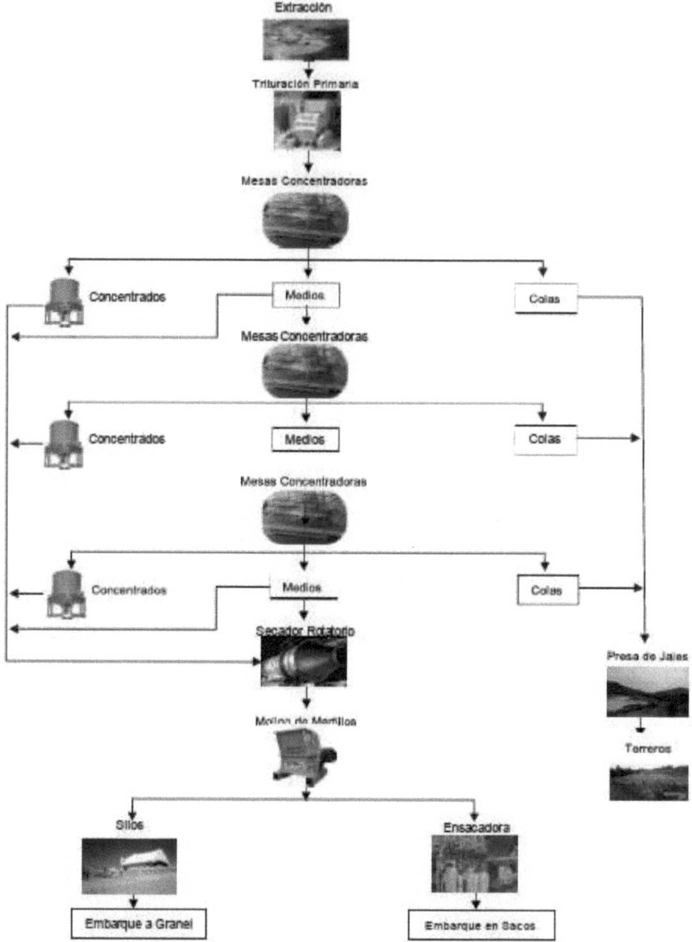

Figure 2.- Barite production process

Extraction

It is carried out by underground mining using the tumbe sobre carga system. Ore extraction is carried out through shafts at haulage levels. In-mine haulage after hoppers is conducted in a mine carriage to the general surface extraction shaft. Only one third of the broken ore is extracted and the rest is left to continue with the block crushing process. The remaining broken ore is extracted until the cuttings reach the upper level.[4] The remaining broken ore is extracted until the cuttings reach the top level.

Crushing

This is done with jaw and/or cone crushers and vibrating screens are usually used in the circuit, partly to maximise crushing efficiency and to reduce ultrafine production. A secondary crushing circuit is also used for the purpose of homogenising particle sizes prior to gravimetric separation.

Concentration

The crushed ore is sorted on concentrator tables where concentrates, media and tails are differentiated, which can be repeated up to three times. The concentrates pass through a rotary dryer, the media through the concentrator tables and finally through the rotary dryer and the tails to the tailings dam and then to the tailings storage area.

Drying

It is carried out by means of a rotary dryer, into which both media and concentrates are fed in order to reduce moisture and ensure the free flow of material in the subsequent steps.

Grinding

The grinding of the ore is done by hammer milling, in order to have a better control of the particle size. On the other hand, grinding is also useful to adjust the size to the customer's needs and specifications.

Packaging

The ground ore is stored in silos to be loaded onto 28-tonne tractor-trailers for bulk shipment, or by a bagging machine to fill 50-kilo bags for subsequent shipment [4].

2.1.4 MAIN USES

Drilling mud raw material

Barium sulphate is a solid that is added to drilling fluids to increase the density in order to prevent gas, oil or water in permeable formations from invading the borehole and to prevent wall collapse by controlling the hydrostatic pressure of the fluid columns, which depends on the density of the added barite and the length of the hydrostatic column.

Paints

Finely ground, it is used in the manufacture of white pigments and as a mineral filler in paints to give body to the pigment. It controls the viscosity of the paint to make brightly coloured products and gives good stability.

Chemistry

The uses of barium carbonate are classified as follows:

2.1.5 Raw material for the production of other barium compounds.

2.1.6 Purification medium for the elimination of all sulphates from aqueous solutions.

2.1.7 Flux in ceramic works.

2.1.8 Ingredient in the manufacture of optical glass and fine glassware.

Medicine

As an opaque medium in tracto-gastrointestinal X-ray examination, for reflection of the intestines and stomach. As a filler for plaster (orthopaedic), used in the production of hydrogen peroxide and in some medicines to extend the Hmite time (shelf life).

Glass

Partially crushed, it is used in furnaces to reduce the melting point of glass.

Filling

As filler in the rubber, leather, rubber, textile and paper industries.

Plastics industry

As a plastic filler to make products with bright colour. It can also improve the strength, stiffness and abrasive resistance. Synthetic barium sulphate helps in the reinforcement of polymers and in the control of rheology and viscosity of adhesives, as well as for the nucleation of crystalline thermoplastics. The fact that it is an inert, temperature resistant product with high transparency and good dispersibility enables it to control the rate and degree of crystallisation.

Automotive applications

For sealing the interior of a vehicle (under carpet) to prevent engine noise, as well as brake linings: pads, discs and brake linings.

Paper industry

Used as a filler for white cardboard and liner paper, it improves whiteness and coverage percentage.

Cosmetics

It is used for its whiteness and gentle and harmless treatment to the skin, so it can be a substitute for titanium dioxide.

Construction

For the production of heavy concretes.

Rubber industry

Products smaller than 500 mesh can be used as filler for rubber products. It can improve the intensity of acid, alkali and water proof; it also has advantage for natural and synthetic rubber.

Coatings

Applicable as fire protection for buildings, airports and gymnasiums. Also on wires, cables, wood, fibreboard, plastics and other flammable substances. Synthetic barium sulphate improves impact resistance, is chemically and mechanically stable, as well as improving geological properties.

Metallurgy

Barium salts are produced from the chemical quality of barite, although witherite is the more important of the two.

Since barite is very insoluble, the starting unit for the manufacture of barium compounds is the reduction of the sulphate ore to barium sulphide: BaS or black ash which is soluble and which by addition of other reagents is converted into a variety of products such as lithopone (a mixture of barium sulphate and zinc sulphide, which is a brilliant white pigment), fixed white ($BaSO_4$), barium chloride ($BaCl_2$), barium nitrate $Ba(NO_3)_2$, barium carbonate $BaCO_3$, barium oxide (BaO), barium dioxide (BaO_2) and finally hydrated barium hydroxide $Ba(OH)_2\ 8H_2O$.

Barium carbonate BaCO3

The uses of barium carbonate fall into several categories:

1) Raw materials for the production of other barium compounds.

2) Purification medium for the removal of all sulphates from aqueous solutions, together with the precipitation of heavy metal ions, alkaline earth metals and magnesium.

3) It is widely used in the ceramics industry as an ingredient in glazes.

4) It acts as a flux, a protective and crystallising agent that combines with certain colouring oxides to produce unique colours not readily achievable by other means.

5) Ingredient in the manufacture of optical glass and fine glassware, due to the optical properties it imparts.

6)   Carbon carrier in the surface hardening baths (this is the type of baths given to new cars at the factory before painting).

The most important industrial use of this material is based on the extreme insolubility of barium sulphate. Although carbonate is considered insoluble in water, it slowly converts to sulphate when agitated with a solution of sulphate ion. If the soluble sulphate is gypsum or magnesium sulphate, the anions and cations are precipitated in turn.

Thus, in the brick industry, the magnesium sulphate and gypsum present in the clay or in the water used to knead it, form a white foam on the surface of the brick when it is dried before firing. This is due to the diffusion of soluble salt to the surface and its crystallisation during drying. If barium carbonate is added to the clay before kneading with water, a reaction with sulphates will occur during kneading and drying, which will prevent the formation of foam on the brick.

Barium sulphate, being the final product of barite metallurgy, is concentrated by flotation and/or gravimetric concentration to the order of 96-98% BaSO4 with a specific gravity of 4.2 to 4.3 gr/cc, which is currently required by the oil industry.

Substitutes

As filler, it can be replaced by calcium carbonate, diatomite, feldspar, kaolin, mica, nepheline syenite, perlite, talc, microcrystalline sHice, fluorsilica, synthetic sHice and wollastonite. In the glass industry it is replaced by strontium carbonate, and as a filler by celestite, hematite, ilmenite and iron ore.

2.1.5 LEGAL FRAMEWORK

Article 4 of the Mining Act specifies that minerals or substances from which barium is extracted, as well as the industrial minerals barite and witherite, are subject to the Mining Act. [5]

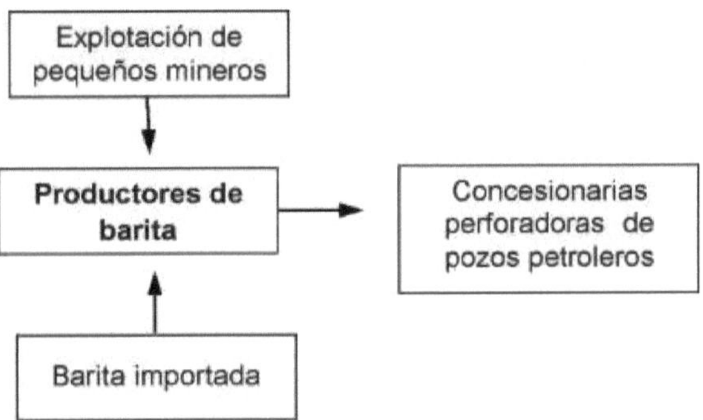

Figure 3.- Baryte marketing scheme

The barite producer sells (in bulk) barium sulphate directly to PEMEX's contractors (oil well drillers). The 'barite producer' takes care of the logistics, and hires the transporter who will take the ore to the points of consumption; whatever the destination of the ore, the price will be fixed.

Some small-scale miners temporarily exploit their deposits and sell the ore to medium-scale miners with their own beneficiation plant, who add it to their production once they have confirmed the quality of the ore, and then transfer it to the transformation process.

As a market strategy, the main barite producers have established plants close to the ports where they receive and process imported barium sulphate in chunks; in this way, they stay close to the consumption points. This reduces freight costs, as their mines are located in the north of the country. The importation of barite is carried out by the national producers themselves as a complement to their production, and they are always in charge of the whole logistical process.

2.2 BACKGROUND INFORMATION ON THE MINERAL CALCITE

Calcite is a mineral consisting of calcium carbonate ($CaCO_3$), of class 05 of the Strunz classification, the so-called carbonate and nitrate minerals. It is sometimes used as a synonym for limestone, although this is incorrect, as limestone is a rock, not a mineral. Its name comes from the Latin *calx*, meaning *quicklime*. It is the most stable calcium carbonate mineral that exists, as opposed to the other two polymorphs with the same chemical formula but different crystalline structure: aragonite and vaterite, which are more unstable and soluble.

Calcite is very common and has a wide distribution throughout the planet, it is estimated that approximately 4 % by weight of the earth's crust is calcite.

It has an enormous variety of shapes and colours. It is characterised by its relatively low hardness (3 on the Mohs scale) and by its high reactivity even with weak acids, such as vinegar, in addition to the aforementioned prominent division into many varieties - hundreds have been described - according to the metal ion impurities it may carry.

The best property to identify calcite is the acid test, as this mineral always effervesces with acids. It can be used as a criterion to determine whether the cement in sandstone and conglomerate rocks is calcite. The reason for this is the following reaction:

$$CaCO_3 + 2H^+ \quad\quad Ca^{2+} + H_2O + CO_2 \text{ (gas)}$$

where the carbon dioxide produces bubbles as it escapes as a gas. Any acid can produce this result, but it is advisable to use dilute hydrochloric acid or vinegar for this test. Other very similar carbonates, such as dolomite, do not produce this reaction as easily.[6] The test can be performed with a dilute hydrochloric acid or vinegar.

It is a mineral occurring in granular or cruciform masses; it grows in cavities, as pyramidal crystals called "dogtooth spar". It is a calcium carbonate ($CaCO_3$), colourless, soft and of various shades, hardness 3, oblique cross-cutting in diamonds (i.e., in three directions); opaque to pearly lustre and a specific gravity of 2.7.

It is a rock-forming mineral and its composition is calcium carbonate ($CaCO_3$), which is equivalent to 44% carbon dioxide and 56% lime. In small amounts magnesium (Mg), iron (Fe), manganese (Mn), zinc (Zn) and lead (Pb) may be present in place of calcium.

Their variation is very numerous and diverse in appearance. It depends mainly on the following points: differences in crystallisation and structural condition, presence of impurities, etc. The extremes being perfect crystals and massive earthy forms. It is used in the manufacture of mortars and cement; as a building and ornamental material, as a flux in metallurgical operations, as pure crystals in the manufacture of optical instruments, lime, fertilisers and lime bleaching.

Calcite, which is a constituent mineral of limestones and marbles, and therefore of cement, is evidenced by its effervescence with official HCl (diluted to 10%).

The availability of high grade limestone is often limited and it is necessary to use low quality sources from quarries located close to cement factories. Beneficiation is necessary to remove impurities and to fix the correct proportion of calcite and other constituents for cement specifications.

Flotation has been successfully applied in the cement industry to improve the composition of finely ground limestone prior to its introduction into the kiln for cement manufacture. In many cases, it is only necessary to divert a part of the mixture through flotation in order to meet the requirements for kiln feed. The flotation process is a means of obtaining a uniform feed and a more uniform end product. Specifications vary, but generally 77-78% $CaCO_3$ is desired in kiln feed.

2.2.1 PROCESSING OF CALCITE

In all cement factories it is necessary to initially crush the calcite-containing rock, and then grind it to the desired fineness for kiln feed.

In some operations the specifications require a size of 90% less than 200 mesh and in others, 100% less than 50 mesh. This procedure follows, but is not necessary even if a flotation stage is not required. The introduction of flotation into an existing plant is not a major investment.

Grinding is generally carried out wet. An upstream analysis of the material treated by flotation is represented by the following [7]:

Figure 4.- Calcite mineral

Calcite (including dolomite)69.5% Calcite (including dolomite)   69.5%
Calcite (including dolomite)
Sflice(quartz) 10.3
Graphitic carbon0 8
Iron (pyrite)..2.5
Mica16.. .7

Calcite Ore Crushing:

In the flowsheet there is a two-stage crushing circuit to reduce the material to be fed to the ball mill to approximately %" in size. The size of the crushing equipment depends largely on the quarrying operation and the maximum size of rock delivered to the primary crusher. This also depends on the mining method used in the quarry and the size of the shovel used to load the trucks transporting the rock to the plant. A 32 "x40" or 36 "x48" jaw crusher is ideal for limestone crushing, and the larger size has a capacity of up to 300 tonnes per hour for a 4" size.

Calcite Ore Grinding:

The crushed ore at minus %" is fed to a ball mill in closed circuit with a classifier to have a size for feeding to the furnaces or flotation. The flowsheet shows a grinding circuit to have a product of 100% minus 50 mesh. For very fine grinding (90% minus 200 mesh), it may be necessary to use a hydroclassifier in closed circuit with the primary classifier and the mill. In such cases it is necessary to thicken the pulp prior to flotation treatment to treat the hydroclassifier overflow.

Calcite delamination:

In many cases it is possible to delaminate and obtain a calcite enrichment in the delaminated fine fraction. This is because calcite is softer than quartz; such a practice is followed in the treatment of certain marine deposits high in lime and silica, but low in other impurities such as graphite and sericitic mica. A hydroclassifier is used for this purpose, although centrifuges and cyclones can also be employed under favourable conditions. If the fine fraction is enriched with calcite, it can be combined with a high grade concentrate obtained from calcite flotation. Extraction of the fine fraction with lamellae often makes the flotation operation more positive and reduces reagent costs.

Calcite flotation

If the ore contains graphitic carbon and/or pyrite, these impurities are floated and discarded as waste. Petroleum or paraffin and a foaming agent such as pine oil or methylated spirits are used to float the graphite. In case pyrite is present and it is desirable to remove it, a xanthate is added to the sulphide flotation and the froth is discarded. The loss of calcite in this product will be very low. Flotation reagents are added to the conditioner and to the flotation cell, keeping them out of the grinding and desliming circuits. The pulp density in the flotation will vary between 20-30% solids, depending on the degree of grinding required to liberate the impurities and to meet furnace feed specifications.

The tailings from the impurity flotation stage are conditioned with fatty acids such as oleic acid and reagent 708, the latter being a vegetable fatty acid. Calcite is activated and floats on the froth while sericite, quartz and other silicates are depressed and rejected in the flotation tailings. The froth from the last 2 or 3 calcite flotation cells can be recirculated if it is desired to maintain the grade and provide a good quality calcite concentrate.

Flotation reagents

In calcite flotation it may be desirable to depress and disperse any graphitic carbon that still exists in the pulp. Calcium lignosulphonate has been used successfully for this purpose.

Saponification of fatty acids with sodium hydroxide and feeding the resulting mixture often improves selectivity on some minerals. A good grade of oleic acid may also be necessary, although in certain cases, a low grade oil or a mixture of different fatty acids and oleic acid have been used with success. The reagents and quantities used are as follows:

Table I. Typical saponification reagents

Areillaceous Limestone	Crustacean Shell Limestone
Petroleum, 0.08 kg/t	Fatty Acid, 0. 44 kg/t
Foaming agent, 010 kg/t	Reagent 708, 0.33 kg/t
Fatty Acid, 0.17 kg/t	Alkali, 0.10 kg/t
Saponified fatty acid, 0.50 kg/t	

The fatty acids are generally saponified and added to the flotation cells in sufficient quantity to give the desired grade of calcite in a concentrate obtained with a flotation time of 10 to 12 minutes.

Thickening and storage

The flotation concentrate with 80% to 85% CaCO3 is thickened together with the hydroclassifier fines if it has a high carbonate content. In some cases, only a portion of these fines may be combined in the flotation concentrate so that the final product has 77% to 78% CaCO3. The discharge from the thickener with 50% to 55% solids is fed to blending tanks for further treatment if necessary before being fed into the kiln to produce cement. The whole process described above is depicted in Figure 5.

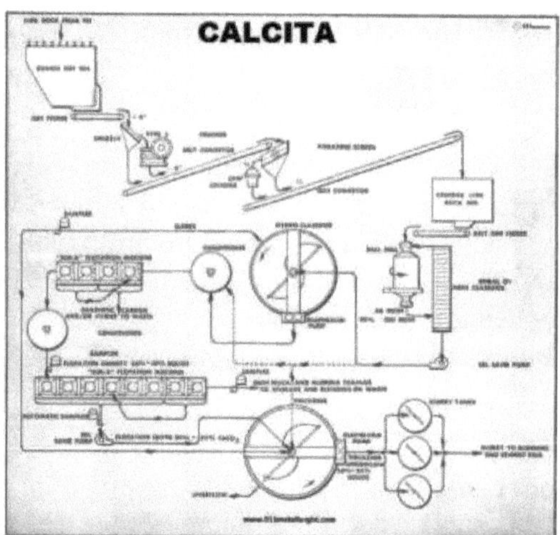

Figure 5. Calcite production process

2.2.2 LOCATION, EXTRACTION AND USES OF CALCITE

It is found in quarries all over the world. It is easily found as associated minerals with siderite, quartz, pyrite, fluorite, dolomite and barite.

It is widely exploited in quarries from which it is extracted in large quantities for a wide variety of uses, from use in the manufacture of cement and mortar, such as limestone and marble building stones, rock and gravel also for construction, agricultural fertilisers for overly acidic soils, or even transparent calcite for the optical industry as microscope polarising prisms. Fused limestone is also used in the metallurgical steel industry and in the manufacture of glass.

Calcite is one of the best type minerals for collecting, as there are many interesting forms and varieties, as well as colourful and beautiful specimens. They are relatively easy to identify by collectors because of their double refraction and acid reaction. However, care must be taken as such a reaction discolours the area which has come into contact with the acid and diminishes the smoothness and lustre, and may even break it down into small pieces of sand. Because of its beauty, calcite has also been used for carving sculptures.

Recently, an invisibility experiment with small objects such as paper clips and pins has been achieved by using two calcite fragments and refracting light.

There are thousands of excellent sites; the original Icelandic Spat variety was described in an old mine on the east coast of Iceland called Helgustadir Mine, in Reydarfjorder, which has been known since 1600.

Two classic European localities where this mineral is found are St. Andreasberg, Harz Mountains, Germany; and Pribram, Bohemia, Czech Republic. Colourless prismatic calcite crystals are well known among collectors from Bigrigg and Egremont, Cumbria, England; and yellow y orange calcite crystals, sometimes in exceptional "butterfly twins", have come from Malmberget, Lappland, Sweden. At present, most of the Iceland Spato on the market comes from Chihuahua, **Mexico**.

Large golden yellow to brown calcite crystals have come from several areas in the Kansas, Missouri and Oklahoma mineral region. These include excellent prismatic crystals from the Sweetwater Mine, Reynolds Co, Missouri; and large scalenohedral crystals from Joplin, Jasper Co, Missouri. Large, lustrous, golden brown crystals from the Elmwood Mine, Carthage, Tennessee.

Calcite with strong blue phosphorescence comes from Terlingua, Brewster Co, Texas. Pink and purple crystals, sometimes very large, have come from Rossie, St. Lawrence Co., New York. White y orange "Salmon Calcite" that has a bright red fluorescence is found in Franklin and Ogdensburg, Sussex Co., New Jersey. Prospect Park, Passaic Co., New Jersey has produced many different types, shapes and colours of this mineral.

<u>Shocking or Crystal Grottoes</u>

In the Mexican state of **Yucatan,** there is an entrance to the "Mayan underworld": the Grutas Chocantes or Grutas de Cristal, one of the largest and most exotic caves in Mexico.

These caves are characterised by their cascades of calcium carbonate, their enormous petrified curtains, their eccentric formations of stalactites, stalagmites and stalagmata. They are named in memory of the owners who discovered them about 65 years ago: the Chocantes.

These Crystal Caves have calcium cascades 6 to 7 metres high and have vaults where the entire floor is white due to the calcite it contains. The INAH and the UNAM have carried out different studies in this place.[8] The caves are also known as "Grutas de Cristal".

Uses of calcite

Calcite is the primary calcium mineral, it is indispensable in the construction industry, forming the basis of cement. It is also crucial in the manufacture of fertilisers, metals, glass, rubber and paint. The transparent Iceland Spar variety, in which the double refraction is very evident, is used as prisms for polarisation microscopes and other optical devices.

This mineral also forms rocks that are used for ornamental purposes, such as marble and banded travertine or tuff. It is also the main component of chalk, which is processed to draw chalk. For collectors, it is one of the best known and most commonly collected minerals. Most sperm specimens are inexpensive, except those of exceptional size and crystal form, or of classic occurrences.

Other uses:

- Animal feed
- Antacid - calcium carbonate
- Baking powder
- Paper manufacturing
- Optical purposes
- Fotograffa
- Statues

Esotericism

It is said in various readings that calcite recharges the chakras, for many, it is the multivitamin for the soul. Composed of calcium carbonate, the Calcite crystal is reputed to be an energy amplifier in a big way. Each variety also has its own distinctive properties that provide the rich and soulful nutrition necessary for creativity and spiritual vision to flourish. Always start the day on the sunny side of life when you take your daily dose of vitamins for the spirit,

The significance of calcite crystal stone can be traced back to ancient Greece, from where the Latin word calc, meaning 'lime', originates. The main ingredient in the shells of sea creatures, crystallised limestone gives it an incredible variety of varieties and colours. Its absorptive powers have made it a powerful sponge that soaks up toxic emotions and negativity.

If your emotional forecast calls for cloudy weather with a chance of rain, insiders recommend always having a little sunshine in your pocket when carrying the Orange Calcite stone. Hold a stone in each hand and give your soul a glowing tan with a dose of vitamin D, the life-sustaining nutrient that radiates with the sun's energy. U

Green calcite

Some say that, in this chaotic world, you can never have too many good luck charms, which makes Green Calcite a must-have stone to invite good fortune and prosperity into your life. Its vibrant emerald hue promotes prosperity and good fortune.

Green is the colour of money, but the cool, soothing tone also symbolises the natural world, and when you meditate with this variety, it reconnects you to the planet as a source of life-giving energy. Every time you look at its soothing green colour, let it remind you of the life-giving earth and its daily abundance. This is the magic of the healing properties of this mineral, especially the green variety. Use this variety to reconnect to the matrix of the universe and remember your immense gratitude for the natural world, a complex system that sustains all life.

2.2.3 PROPERTIES OF CALCITES

Physical Properties

Calcite is colourless or white when pure, but can be almost any colour: reddish, pink, yellow, yellow, greenish, bluish, lavender, black or brown, due to the presence of various impurities. It can be transparent, translucent or opaque.

Its lustre varies from glassy to opaque; many crystals, especially colourless ones, are glassy, while granular masses, especially fine-grained ones, tend to be opaque. Calcite is number 3 on the Mohs scale of hardness; it can be easily scratched with a razor or geological pick.

It has a specific gravity of 2.71. Three perfect divisions give calcite its six-sided polyhedra with diamond-shaped faces; the defining angles of the faces are 78 and 102 degrees.

<u>Optical properties</u>

- Optical properties: transparent to opaque.
- Colour: colourless or white, also grey, yellow, green, many other mineral colours included; colourless in transmitted light.
- Lustre: vftreo; pearlescent on necklines and {0001}.
- Optical class: Uniaxial (-); Anomalously biaxial.
- Dispersion: Very strong.
- Absorption: 0>E. w= 1.658 £ = 1.486 2V (measured) = 0 ° at small [9].

2.3 . GENERAL CHARACTERISTICS OF FLUORSPAR ORE

Fluorite or fluorospar is chemically a calcium fluoride (CaF_2), it belongs to the chemical class of halides, which is characterised by the preponderance of electronegative halogens such as Cl-, Br-, F- or I-. It has a vitreous lustre; its specific gravity is between 3 and 3.2; it has a hardness of 4; it varies from colourless to shades of blue, purple, green and yellow, among others; it is fragile and presents perfect exfoliation. It occurs in very well formed cubic crystals, often with penetrative twins. On the other hand, it is frequently fluorescent and phosphorescent when heated or exposed to radiation. It usually contains mineral impurities such as calcite, quartz, barite, celestite, sulphides and phosphates. Commercial fluorspar is classified according to quality and specification into acid, metallurgical and ceramic grades, thus the grade determines its end use. It is worth mentioning that the uses of fluorspar in order of importance are mainly focused on the production of hydrofluoric acid, steel fabrication and aluminium production. When sulphuric acid is added, it decomposes into hydrogen fluoride gas and calcium sulphate, which is the fundamental reaction to produce hydrofluoric acid; it provides fluidity at low temperatures.[10-12] Fluorite is also used in the production of aluminium.

2.3.1 VARIETIES SPECIFICATIONS

Acid Grade
Very fine concentrate, with a minimum content of 97.1% calcium fluoride (CaF_2). Impurities included are: <1.5% SiO_2, 0.03-0.1% sulphur as sulphur and/or free. Other limitations include: <10-12 ppm arsenic, average phosphorus between 100 and 550 ppm. Total restriction for

lead, cadmium, beryl, calcium carbonate and mixtures. Particle size is -100 mesh (product of flotation)[13].

Ceramic Grade

There are three types of ceramic grade, varying from 85 to 96% CaF2, with the following limits for impurities, which must be kept to less than 2.5-3.0% SiO2, 0.12% ferric oxide, very limited limestone, lead and zinc sulphate, only trace amounts.

Metallurgical Grade

Contains an effective minimum of 60% fluorspar, as impurities: <0.3% sulphide, <0.5% lead. Calcium fluoride content can range from 70% to 93% with a maximum of 15% SiO2, usually required to pass through a 1-1.5 inch mesh (can be up to -8 inches depending on customer requirements), containing less than 15% of material smaller than 1/16 inch [14,15].

2.3.2 Geological Mining Potential

The fluorspar producing states in our country are San Luis Potos^ (93 %) and Coahuila (7%), producing 100.0 % of the national total. The deposits in San Luis Potos^ correspond to large bodies formed in the contact of carbonate rocks with tertiary volcanic rocks. Structurally, mineralisation is associated with the development of a karstic structure where deposition is due to open spaces and replacement of calcareous strata. The Las Cuevas deposit is one of the largest high grade fluorite deposits, consisting of a group of massive bodies encased in contact zones between the El Doctor Formation and a Tertiary rhyolite breccia, which is the deposit wall, consisting of rhyolite clasts in a clayey matrix, is highly altered with small fluorite veinlets filling open spaces in the breccia, indicating that this unit predated the mineralisation. The fluorite body itself is brecciated with fluorite and limestone forming clasts in a fluorite and calcite cementitite. Several breccia types exist in the deposit and exhibit a systematic distribution. The ore grade varies from 50% to 98.5% CaF2, these values depend on the gangue minerals they contain, quartz and calcite being the main contaminants. In Coahuila, the deposits are mostly lenses and replacement mantos, which are presented in a configuration of linked lenses of variable dimensions. There are some chimney-shaped deposits. [16-19]

Figure 6. Map of Mexico with main Fluorspar industries

In the La Linda - Aguachile zone, in the municipality of Acuna, Coah., the fluorite mines of Aguachile and Cuatro Palmas are located, where the mineralization is manifested in replacement bodies, mantos, disseminates and large chimneys. The fluorite is associated with beryl, which suggests a close relationship between the fluorite-bearing chimneys and the alkaline igneous rocks exposed in the area. The deposits are located in limestone of the Santa Elena Formation. Deposits in the Buenavista zone mostly occur as mantos and very occasionally as veins. Thicknesses vary from 0.6 to 2 m and have distributions of several hundred m2 making it one of the districts with the greatest mining potential. The host rock of these deposits is reef limestone of the Lower Cretaceous Georgetown Formation and shale of the Upper Cretaceous Del Rfo Formation. The contact between the two formations is a prospecting target as this is where most of the reservoirs are located. It is worth mentioning that there are other localities in different states with fluorite occurrences such as Zacualpan, Edo. Mex., Galeana NL, San Francisco del Oro, Chih. and in Zacatecas in areas such as Jalpa, Huanusco, Tabasco, Tayahua, La Blanca, Villa Hidalgo and Estacion Frio.[20] The fluorite is also found in the Zacatecas region of Mexico.

2.3.3 Production Process

Dehydration Sampling

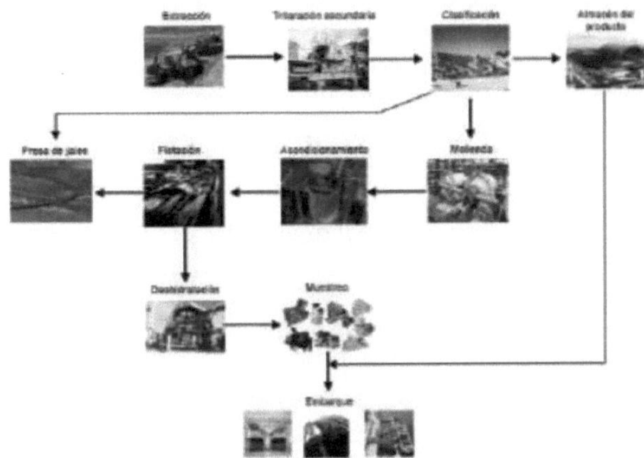

Figure 7. Fluorspar production process

Extraction

The deposits are mined by the sub-level stoping method, which consists of stoping the ore between levels, the distance between which varies. Drilling is carried out with special equipment for semi-long and long holes, the length of which varies from 7.5 to 20.0 metres. The diameter of the holes ranges from 2 to 5 inches. The extraction of the ore is carried out through crossings located on the level below the sublevels and through which the ore is transported by trucks from loading stations to the primary crushing plant, where the large rock fragments are crushed to 8 inches, both activities are carried out inside the mine. This is followed by the handling stage, which consists of extracting the ore from inside the mine to the surface by means of special shafts, i.e., extracting the ore in extraction boats.[21] The ore is then extracted from the mine by means of a special shaft, which is used for this purpose.

Secondary crushing

The ore is then discharged into a hopper to be transported by a winch to the surface, where it is conveyed by means of a belt system to the secondary crusher, where the grain size obtained in the previous process is further reduced, resulting in a product with a size of 21/4 inches.

Classification

A screening system separates the metallurgical grade ore (equal to or greater than % of an inch), the fines (less than % of an inch) are sent to the beneficiation plant and the fines to a stockpile for onward shipment to customers. Impurities such as calcium carbonate and sHice are sent to the tailings dam[22-25].

Grinding
The mineral intergrowth and the fine dissemination of the mass that forms the rock make it necessary to pulverise the ore. This process is carried out by means of ball mills, resulting in a pulp with a grain size of 325 mesh as the final product of this process.

Conditioning
The grinding product is sent to the conditioning tanks where the most important thing is to add reagent solutions to achieve a more efficient separation of the minerals.

Flotation
The pulp obtained in the milling stage is sent directly to a bank of flotation cells for concentration, where air is injected to generate bubbles, which together with the chemical reagents cause the fluorspar to float and the precipitation of silicon and calcium carbonate to take place, which are sent to the tailings dam. The fluorspar is then sent to a thickener tank to remove the water and obtain the concentrate with a minimum content of 97.3% $CaF2$.

Dehydration
The product obtained in the previous stage passes from the thickener to the filters that eliminate the water, leaving the concentrate with a maximum moisture content of 10%, the mineral is retained in the filters, thus obtaining the final product that is sent to the warehouse.

Sampling
Before the fluorspar concentrate is delivered to the users, sampling and chemical analysis is carried out in the laboratory to check the content of the main components (calcium fluoride, calcium carbonate, silicon oxide) and to meet the customers' specifications.

Embarkation
Most of the final products are shipped to the chemical (acid grade) and steel (metallurgical grade) industries in bulk, the means of transport being rail or trailer and ship. Occasionally, it is shipped in supersacks to the

2.3.4. Hydrofluoric acid process

Acid grade fluorspar is the most common and is the raw material for the production of fluor^dic acid. Its process is as follows [26-29]:

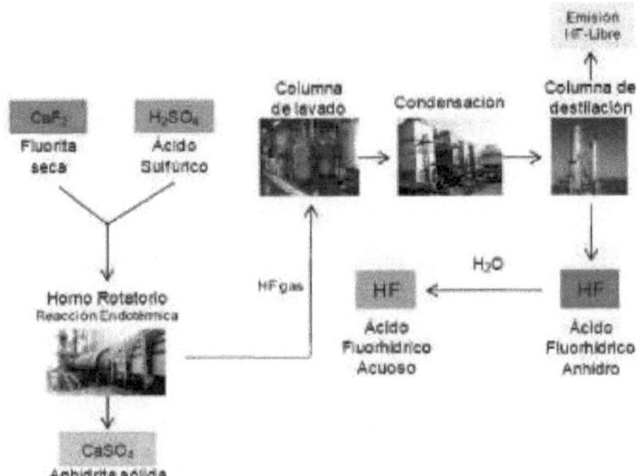

Figure 8. Acid grade fluorspar production process

Fluor^dic acid is produced from acid-grade fluorspar which is reacted with sulphuric acid in a rotary kiln to produce hydrogen fluoride gas. Acid grade fluorspar typically contains at least 97% calcium fluoride, as well as sHice, calcium carbonate, carbon, sulphur, phosphorus pentoxide, chloride, mixed metal oxides and traces of arsenic. The resulting residue after calcination is anhydrite (calcium sulphate without water), commonly known as fluorogypsum, which is a suspended solid waste removed to waste management. The hydrogen fluoride gas is sent to a scrubbing column where it is cooled and washed, then condensed and passed through a distillation process to remove remaining water and impurities and sold as anhydrous hydrogen fluoride. The product can be diluted and sold as an acidic fluorohydrogen fluoride solution of approximately 70%. The waste from the refrigerated condenser goes to an acid scrubber. The sulphuric acid used in this process unit is sent to the acid feed, to react with the fresh fluorspar. The stream from the acid scrubber is sent to a water scrubber which generates the acid.

2.3.5 Main Uses

Acid Grade

It is a base material in the production of fluorhydric acid, which is the starting point for a broad spectrum of fluoride-based chemicals.

<u>Fluorides</u>

Aluminium polishing; electrolytic bath additive; separation of niobium and tantalum; etching or chemical corrosion of printed circuit boards. As a component in aluminium refining; fluxing and pre-oxidising agent in enamels; component of coatings on electrodes and welding powders. As a flux in non-ferrous metals, in zirconium and titanium brazing and as a constituent of aluminium solder. In insecticides; matte or etched glass; battery electrolyte; fluorination of organic compounds by halogen exchange; catalysis in the manufacture of polyurethane foam; ingredient in cement coatings, electric arc welding rolls; table salt processing; ingredient in welding fluxes; pretreatment formulations used to prepare metal surfaces for painting; photographic colour transfer process; base or elemental material in the production of fluorine, sodium fluoroacetate and hexafluorobenzene. In water fluorination; manufacture of organofluor compounds; insecticides; metallurgical flux; wood preservation; satin finish for aluminium; glass etching; welding powder constituent; denUfric additive; sterilisation in beverages and dairy equipment. As an organofluoride for flame retardancy; as a grease and dirt repellent in textiles; emulsifier for tetrafluoroethane polymerisation; for flow control in paints. Sodium fluoride is used in water fluorination, wood preservation, dental additive and sterilisation of beverage equipment [30].

Metallurgical Grade

In steelmaking to reduce slag surface tension, minimise variations in slag viscosity with melting temperature, obtain the lowest melting point of slag to form a eutectic (flux), improve slag fluidity and heat transfer (slag removes Si, S, P and C from iron). It is used in electric, basic oxygen and open-hearth furnaces, as well as in iron foundries. In uses other than steelmaking, it is used as a binding agent for grinding wheels, in the production of carbide and calcium cyanamide. Minor uses include ornamental purposes such as carved vases, ornaments and figurines. It functions as a catalyst, saves energy and lowers the melting temperature (between 200 and 300°C) in iron, steel and cement. In metallurgical grade fluorspar there is a subgroup called cement grade fluorspar which is used as a catalyst, impurity cleaner and flux lowering the melting temperature in the clinkerisation process, also, one of the main benefits is the energy saving as^ well as the quality improvement in the final product. [31]

Ceramic Grade

Opacifier in opal and flint crystals for the food, beverage, toiletry and ornamental glassware industries; as well as in enamels for cookers, refrigerators, bathroom pipes, cabinets and cooking utensils. It is an ingredient in the manufacture of magnesium and calcium, manganese chemicals, zinc die casting, fibreglass and coating of welding rods. Substitutes Chlorofluorocarbons (CFCs) 11, 12 and 113, hydrochlorofluorocarbons (HCFCs) 22, 123, 124, 141b, 142b and 225, hydrofluorocarbons (HFCs). In smelting, borates, dolomitic lime, ilmenite, manganese, mixtures of iron, aluminium, calcium oxide, olivine and sodium carbonate. Fluorosilicic acid, a by-product of phosphoric acid production, is a substitute in the production of aluminium fluoride and possibly in the production of fluorhydnic acid. [32-34]

2.4 BACKGROUND INFORMATION ON MOLYBDENITE ORE

2.4.1 Historical Background

The word molybdenum comes from the Greek word "molibdos", meaning lead, and was used to denote substances which, according to common understanding, were lead, but which in addition to lead could include galena, molybdenite and graphite.

The first clarifications began in the second part of the 18th century, when it was shown that both molybdenite[5] and graphite do not contain lead. Then, in 1778, C.W. Scheele showed that molybdenite can be decomposed by attack with nitric acid, producing a white powder with acidic properties which he called "molybdic acid".

Four years later, in 1782, P. J. Hjelm succeeded in reducing molybdic oxide with carbon, obtaining a dark grey metallic powder with metallic properties, which he called "molybdenum". Thus, for the first time, the new element was clearly recognised: molybdenum metal [35].

Molybdenum remained largely a laboratory curiosity for most of the 19th century until the technology for the extraction of commercial quantities was achieved. In 1891, the French company Schneider & Co. was the first to use molybdenum as an element in alloys for the production of armour. It was soon realised that, with a density only slightly more than half that of tungsten, molybdenum was an effective substitute for tungsten in numerous steel alloy applications [36].

2.4.2 Properties of molybdenum

Molybdenum is a metal with atomic number 42, atomic weight 95.94 and is symbolised as Mo. It belongs to the sixth group of the Periodic Table between chromium and tungsten. Although molybdenum is sometimes described as a "heavy metal", its properties are very different from those of the typical heavy metals, mercury, thallium and lead. Its low toxicity makes molybdenum an attractive substitute for toxic metals in numerous applications, for example in place of chromium in corrosion inhibitors and antimony in smoke suppressants [36].

Among the properties of molybdenum, technically interesting for practical uses, it is worth mentioning its high melting point, its extraordinary properties as an alloying element, its ability to easily change valency and, as a negative point, its susceptibility to oxidation at high temperatures.

Of great importance is the mechanical strength of molybdenum steels at elevated temperatures. It has been noted that small amounts of molybdenum eliminate the brittle properties which appear in steels at high temperatures, between 450 and 600 °C. This gives rise to high-speed steels which are used for the manufacture of high-speed drilling, polishing or cutting elements [35].

2.4.3 Molybdenite processing

Molybdenum lies in its deposits in the form of low-grade ores and, before being used for metallurgical and industrial purposes, it necessarily has to undergo beneficiation and concentration operations, which liberate the molybdenum ores from the associated mineralogical species and increase the molybdenum content. The end product of an ore beneficiation plant is the high-grade concentrate, which contains the mineralogical species almost in its pure state. From the concentrate, molybdenum is obtained by pyro-metallurgical, hydro-metallurgical and electrometallurgical processes in its pure metallic form or in the form of compounds that are used in industry.

Molybdenite, the main and almost exclusive source of molybdenum, due to its hexagonal crystalline structure, lamellar formation and strongly hydrophobic properties, is very susceptible to concentration processes by means of flotation.

Another molybdenum ore susceptible to concentration processes is wulfenite, lead molybdate. However, the final recovery of molybdenum involves smelting, leaching and precipitation operations to separate it from lead and other metals. This is why wulfenite is hardly ever used as a source of molybdenum.

Other molybdenum ores, particularly powelite[1] and molybdite[1] , are rather soft yellowish powders with the regular properties of all oxidised ores and, for this reason, do not have favourable conditions for mechanical preparation and concentration.

In summary, it can be considered that almost all molybdenum comes from molybdenite, and that this mineral is exclusively beneficiated by the flotation method [35].

The Mo content of viable deposits varies between 0.01% and 0.01%. and 0.25%, and is generally associated with sulphide ores of other metals, notably copper.

The deposits and mines from which molybdenum is extracted are classified into three types:

•   Mines where molybdenite recovery is the only objective
•   Mines in which molybdenite is separated during copper recovery
•   Mines where the recovery of the co-products molybdenite and copper depends on their commercial viability [36].

The following will explain the beneficiation process for the case where the separation of molybdenum is carried out during the recovery of copper in the copper extraction process. As in all ore beneficiation operations, three stages are essential:
1. liberation of minerals by means of size reduction and classification operations;
2. separation of the noble part from the gangue, or concentration, by means of flotation; and
3. water removal, filtration and drying operations to obtain molybdenum concentrate from its slurries to a commercial dry product [35].

Figure 9 shows in schematic form the main processes involved in the production of molybdenum products.

The molybdenite beneficiation stages are carried out by the following operations:

Grinding: the ore is crushed and ground in ball or roller mills to fine particles with a size below 150 microns.

Flotation: is carried out in aerated tanks to separate metallic minerals from gangue (worthless ore) and - in the case of copper/molybdenum ores - to separate molybdenite from copper sulphide. The resulting molybdenite concentrate contains between 85% and 92% MoS2. If required, acid leaching can be used to dissolve impurities such as copper and lead.

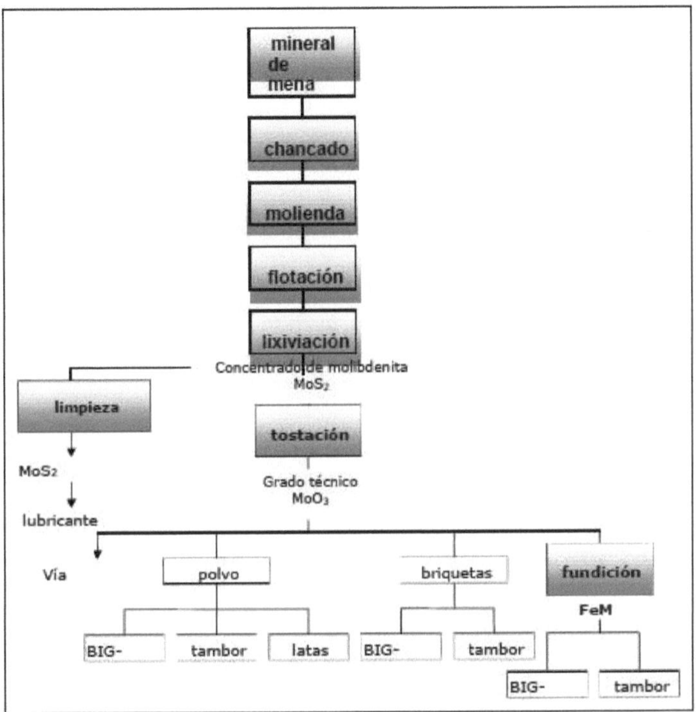

Figure 9: Molybdenite processing scheme Source: Reference [35]

Roasting: this process converts molybdenite into roasted molybdenite concentrate (also known as technical grade Mo oxide) at temperatures between 500 and 650°C:

Roasters are multi-stage shaft furnaces, in which the molybdenite concentrate is moved from the top to the bottom against the stream of hot air and gases flowing from below. The molybdenite concentrate is moved by large rakes to promote chemical reactions. The resulting molybdenite concentrate contains typically a minimum of 57% Mo, and less than 0.1% sulphide.

Rhenium recovery: Some of the molybdenite concentrates that are by-products of copper mines contain small amounts (< 0.1%) of rhenium. Molybdenum roasters equipped for rhenium recovery are one of the main sources for this rare metal.

Ferro-molybdenum smelting: 30-40% of the roasted molybdenite concentrate production is further processed to produce ferromolybdenum. The roasted concentrate is mixed with iron oxide and reduced with aluminium in a thermal reaction, producing a bar of several hundred kilograms of ferro molybdenum containing 60-70% molybdenum and the rest essentially steel. After air cooling, the bar is crushed and sorted through screens to achieve the specified size ranges for the ferro molybdenum product.

Purification of roasted molybdenite concentrates (technical grade Mo oxide): about 25% of the technical grade molybdenum oxide produced worldwide is further processed into a range of chemical products. Purification is carried out:

*   By sublimation to produce pure molybdenum oxide (MoOa)
*   By wet chemical processes to produce a wide range of
molybdenum chemicals (mainly molybdenum oxides and molybdates)
The latter involves initial dissolution of the technical grade oxide in an alkaline medium (ammonium or sodium hydroxide[5]), followed by removal of impurities by precipitation and filtration and/or solvent extraction. The resulting ammonium molybdate solution is then converted into one of the molybdate products by crystallisation or acid precipitation. These can be further processed by calcination to produce pure molybdenum trioxide.

Production of metallic molybdenum: metallic molybdenum is produced by the reduction with hydrogen of pure molybdenum oxide or ammonium molybdates [36].

2.4.4 Main molybdenum products

The main molybdenum products traded internationally are:

a.   Molybdenite concentrate
b.   Molybdenum trioxide, technical and pure grades
c.   Ferromolybdenum
d.   Molybdenum salts
e.   Molybdenum metal

The main characteristics of these products are briefly outlined below:

Molybdenite concentrate

Molybdenite concentrate is the second most traded of all molybdenum products (after technical grade oxide), as it already possesses basic purity conditions and is suitable for transformation into other products. It is the general policy of many companies to buy this product as raw material and transform it into what they need.

The US Government specifications, which are at the same time the minimum conditions on the international market, indicate a product having the compositions shown in the following table.

Table II: Molybdenite concentrate specifications.

Element or compound	Maximum content (%)
Molybdenum sulphide	80,00
Copper	1,00
Lead	0,30
Phosphorus, stannous, arsenic	0,20

Source: Reference [35]

However, it is common to find much higher grade molybdenite products on the market. Generally a product can be obtained that has about 90% MoS2 and less than 0.5% copper and is almost free of stannous, arsenic and phosphorus. The molybdenite concentration in commercial concentrates can be as high as 96% MoS2.

Molybdenite concentrates, which are used for the production of lubricants, must necessarily undergo special purification in order to free them from the abrasive part of their impurities, particularly silica. Various methods are used for this purpose, the most common being leaching with hydrofluoric acid. The final product has more than 99.9% MoS2 [35].

Technical grade molybdenum trioxide

Technical grade molybdenum trioxide is the most widely consumed product. The specifications of this product are shown in Table III.

Table III: Technical grade molybdenum trioxide specifications.

Element or compound	Maximum content (%)
Molybdenum	80,00
Copper	1,00
Sulphur	0,25
Phosphorus	0,05

Source: Reference [35]

As the roasting product contains all the non-volatile impurities in the molybdenite concentrate, it must be purified by sublimation to obtain a special-purpose product. The pure product has more than 99.5% molphodic oxide (MoO_3).

Ferromolybdenum

This product is a very important part of the national consumption of molybdenum, and it is produced in two types: type A and type B. In both cases the molybdenum content is between 60 and 70%, and the maximum impurity content is: 1% copper, 0.1% phosphorus, 0.25% sulphur and 1.5% silica. The difference between type A and B is that type A allows up to 2.5% carbon and type B up to 0.25%.

Molybdenum salts

Molybdenum salts for sale are mainly calcium, sodium and ammonium molybdates as well as molybdenum silicate and various other products and reagents. Sodium molybdate ($Na_2MoO_4 \cdot 2H_2O$) has a 39% molybdenum content, while ammonium dimolybdate [$(NH_4)_2Mo_2O_7$] has 56.4% molybdenum.

Molybdenum metal

This product can be prepared to a purity of 99.95 Mo, but for technical uses a powder having the specifications shown in the following Table is acceptable.

Table IV: Metal molybdenum trioxide specifications.

Element or compound	Maximum content
	95,0)0
Molybdenum	
Fierro	2,50
Silica	1,50
Copper	0,50
Sulphur	0,10
Carbon	0,05
Phosphorus	0,04

Source: Reference [35]

2.4.5 Uses of molybdenum

The main uses of molybdenum are in metallurgy: here molybdenum is used as an alloying element and as a metallic molybdenum. This metal also plays a vital role in environmental and human health protection, as its compounds are intrinsically safe and non-toxic. Hence its use in lubricants, catalysts, corrosion inhibitors and smoke suppressants. The main metallurgical and chemical uses of molybdenum are detailed below.

2.4.6 Molybdenum applications in metallurgy

<u>Molybdenum-based alloys</u>

Due to its high melting point, molybdenum metal has excellent fracture toughness in reducing and high temperature environments. To further increase its strength, fracture toughness and maximise its operating temperature, elements such as titanium, zirconium, tungsten and rhenium can be added to molybdenum.

Molybdenum and its alloys are also used in a wide variety of applications for their thermal and electrical conductivity, thermal expansion rate, stability in different environments and resistance to abrasion and wear.

Molybdenum metal is usually produced through powder metallurgy techniques in which molybdenum powder is compacted and sintered at a temperature of approximately 2,100 °C.

Molybdenum alloys have excellent strength and mechanical stability at high temperatures (above 1,900 °C). Their high ductility and toughness provide better tolerance to flaws and brittle fracture than ceramic materials.

<u>Molybdenum in stainless steels</u>

Additions of molybdenum above 7% strongly improve the corrosion resistance of stainless steel in a wide range of environments. This is especially effective in improving corrosion resistance in environments containing chlorides. Molybdenum also increases the resistance of stainless steel to high temperatures and fracture.

<u>Molybdenum in steel and iron alloys</u>

Molybdenum is used efficiently and economically in steel and iron alloys for:

- Improve its hardness
- Resist hydrogen attack
- Increase resistance to high temperatures
- Improve its weldability, especially in high strength low alloy steels (HSLA).

End uses cover a wide variety of engineered products for:

- Automotive, marine and aerospace
- Drilling and mining processes
- Power generation, including boilers, steam turbines and electric generators
- Reactors, tanks, heat exchangers
- Chemical and petrochemical processes

In most cases only small amounts of molybdenum need to be added. In fact, with the exception of high-speed steel, the molybdenum content is generally between 0.2 and 0.5 % and rarely exceeds 1 %.

<u>Molybdenum in superalloys and nickel-based alloys</u>

In age-hardenable nickel-based superalloys, which are used for high temperature applications such as turbines, the addition of molybdenum above 10% increases the high temperature resistance of the base metal.

In corrosion resistant nickel based alloys, molybdenum (usually about 16%, in special cases even 28%) improves corrosion resistance and mechanical properties.

2.4.7 Chemical applications of molybdenum [36].

<u>Catalytic converters</u>

Molybdenum-based catalysts have a number of important applications in the petroleum and plastics industries. The major use is in the hydrodesulphurisation (HDS) of petroleum, petrochemicals and coal-derived fuels. The catalyst consists of MoS_2 supported on aluminium and promoted by cobalt or nickel, and is prepared by sulphurising cobalt and molybdenum oxides in alumina. As the world's crude oil supply is widely extended and low sulphur crudes are becoming increasingly scarce, the use of molybdenum-based catalysts will increase. Molybdenum not only enables economical refining of fuels, but also contributes to a cleaner environment through lower sulphur emissions.

Molybdenum catalysts are resistant to sulphur poisoning and, for example, catalyse the conversion of hydrogen and carbon monoxide from the pyrolysis of waste materials to alcohols in the presence of sulphides, under conditions that would poison precious metal catalysts. Similarly, molybdenum-based catalysts have been used in the conversion of carbon to kichid hydrocarbons. As a component of the bismuth molybdate catalyst for selective oxidations, molybdenum is involved in the selective oxidation of, for example, propane, ammonia and air to acrylonitrile, acetonitrile and other chemical compounds corresponding to raw materials for the plastics and fibre industries. Similarly, molybdenum in iron molybdate catalyses the selective oxidation of methanol to formalde^de.

<u>Pigments</u>

Molybdate-based pigments are used for two properties: stable colour formation and corrosion inhibition. Molybdenum orange is prepared by co-precipitation of lead chromate, lead molybdate and lead sulphate. These are light, heat-stable pigments with colours ranging from bright red-orange to yellowish-red and are used in paints and inks, plastic and rubber products, and ceramics.

Zinc molybdate is the basis for the corrosion inhibiting white pigments that are used to prepare paints. Molybdophosphoric acid is used to precipitate violet and victoria blue inks.

<u>Corrosion inhibitors</u>

Sodium molybdate has been used for many years as a substitute for chromates as corrosion inhibitors on light steels over a wide pH range. Molybdates have a very low toxicity and are less aggressive oxidants than chromates with respect to the organic additives to be used in corrosion inhibitor formulations.

The major application is in the cooling water of air conditioning and heating systems to protect the light steels used in their construction.

Molybdates are used to inhibit corrosion in hydraulic equipment and automotive engine antifreeze. Molybdate solutions are used to protect machine parts made of steel against rust.

Corrosion inhibiting pigments, mainly zinc molybdates, but also calcium molybdates, are used commercially in paints. These pigments are white and can be used as a base or as an ink with any other colour.

<u>Smoke suppressants</u>

In electronics technology, the insulation of wires and cables represents a potential fire and smoke hazard in the confined spaces of aircraft and hospitals. Ammonium octamolybdate is used with PVC to suppress smoke formation. Its uses and other developments will increase, due to the growth of video, telephone and computer networks.

<u>Lubricants</u>

Molybdenum disulphide (MoS_2), the most common natural form of molybdenum, is extracted from the ore and then purified for direct use in lubricants. Molybdenum disulphide, thanks to its layered structure, is a very effective lubricant. When MoS_2 particles are located between moving surfaces, the MoS2 layers slide past each other, allowing the surfaces of steel or other metals to move smoothly, even under severe pressures. Because MoS2 is of geothermal origin, it has the durability to withstand heat and pressure. This is particularly the case if small amounts of sulphur are available to react with the steel and provide a sulphur layer that is compatible with MoS2 to maintain the lubricating film.

A combination of water-soluble molybdates and sulphides can provide both lubrication and corrosion inhibition in metal-to-metal fluids. Oil compounds with soluble sulphides and molybdenum, such as thiophosphates and thiocarbamates, provide engine protection against wear, oxidation and corrosion. Various commercial manufacturers offer these additives to the lubricant industry.

<u>Molybdenum chemicals in agriculture</u>

Molybdenum is an essential element - in plants, in enzymes that catalyse nitrogen fixation (nitrogenase) and nitrate reduction (nitrate reductase); in animals, in enzymes involved in, for example, nitrogen and sulphur metabolism. Some soils, especially acid soils, require supplementary molybdenum for healthy plant life. For humans molybdenum is a component of a number of vitamin and mineral supplements. Molybdenum is generally classified as non-toxic to humans [37].

Molybdenum's vital role in biology derives from its affinity for sulphur, its finely balanced oxidation states, and its ability in its highest oxidation state to transfer oxygen atoms.

The main uses of molybdenum, together with the processes of its conversion from technical grade trioxide, are shown in the figure below.

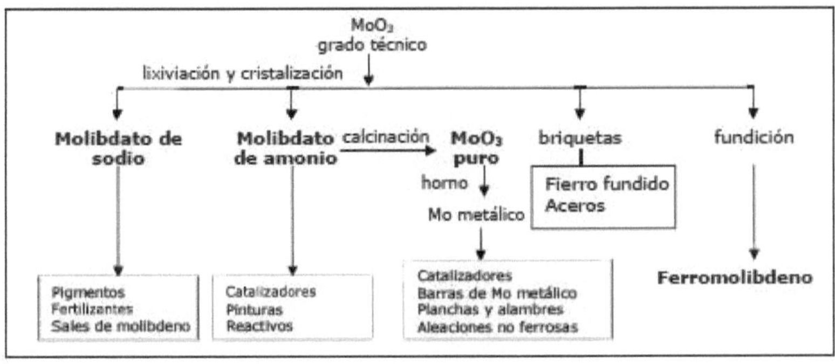

2.5 COCONUT SHELL ACTIVATED CARBON

2.5.1 -Concept Carbon Assets

Activated carbon is a product that has a lattice crystalline structure similar to graphite; it is extremely porous, and can develop surface areas of up to 1,500 square metres per gram of carbon.

All the carbon atoms on the surface of a crystal are capable of attracting molecules of compounds that cause undesirable colour, odour or taste; the difference with an activated carbon is the amount of surface atoms available for adsorption. In other words, the activation of any carbon consists of multiplying the surface area by creating a porous structure. It is important to mention that the surface area of activated carbon is internal. To give us a clearer idea of the magnitude of it, let us imagine a gram of lump carbon which we grind very finely to increase its surface area, as a result we will obtain an area of approximately 3 to 4 square metres, on the other hand, by activating the carbon we achieve a multiplication of 200 to 300 times this value [38].

Therefore, when it is desired to remove an organic impurity that causes undesirable colour, odour or taste, adsorption with activated carbon is usually the most economical and simplest technique.

Activated carbon is a general term for a range of products derived from carbonaceous materials. It is a material that has an exceptionally high surface area.

The name activated carbon is applied to a series of porous carbons artificially prepared through a carbonisation process to exhibit a high degree of porosity and a high internal surface area.

It is a product obtained from amorphous carbon, which has undergone an activation treatment in order to increase its surface area up to 300 times due to the formation of internal pores, reaching areas of 1200 - 1500 m^2 /gr of carbon [31].

The fundamental difference between the two types of carbon lies in the structure, or arrangement of its atoms. In the case of activated carbon, these are combined in the form of graphitic plates, which can be represented according to Figure 11.

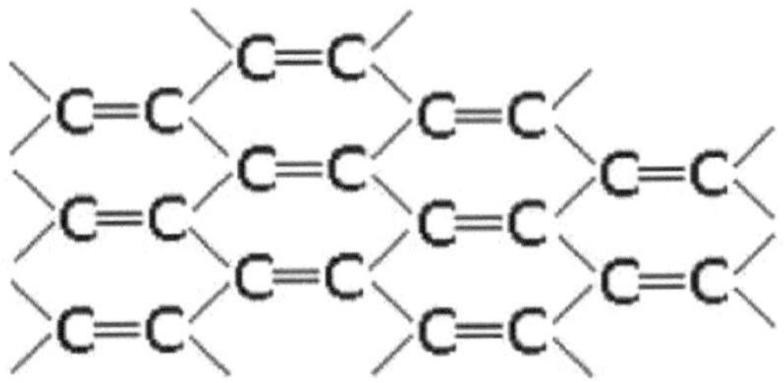

Figure 11. Arrangement of the plates in the activated carbon

The plates are separated and have different orientations, so there are spaces between them, which are called pores, which give the activated carbon its main characteristic: a large surface area, and therefore a high adsorption capacity. The surface area of most commercial activated carbons is between 500 - 1500 m^2 /gr.

Activated carbon is a carbon material that is prepared in industry to have a high internal surface area in order to adsorb a large number of different compounds on its surface, both in the gaseous phase and in solution.

It is a porous material obtained by carbonisation and activation of organic materials, especially of vegetable origin, coal, lignite and peat, in order to obtain a high degree of porosity and a large intraparticle surface area. The high specific surface area facilitates the physical adsorption of gases and vapours of gaseous mixtures or substances dispersed in liquids.

The activity of carbon in adsorption processes is mainly determined by the nature of the raw material and the activation process used in its production. Its adsorptive capacity is greatly favoured by the porous structure and the interaction with polar and non-polar adsorbates, given its chemical structure, and the chemical reactions on its surface are influenced by active centres, dislocations and discontinuities, where the carbons have unpaired electrons and unsaturated valences presenting higher potential energy.

Based on X-ray diffraction, two types of porous structures can be described for activated carbon:

- The first type of structure is formed by microcrystallites which, in two planes, are similar to graphite made up of parallel layers of carbon atoms, arranged hexagonally.

- The second type of structure is described as a three-dimensional lattice of disordered carbon hexagons resulting from the random ordering of condensed benzene structures formed during carbonisation.

Between the microcrystals that constitute the carbon, there are voids or watt spaces which are called pores. The total surface area of these pores, including the area of their walls, is very large, which is the main reason for their high absorption capacity.

The generic term carbon is used for compounds with the regularly arranged element carbon (C) in their composition. The atoms located in the outer part of the crystalline structure have free attractive forces, which allow them to attract compounds in their immediate vicinity [40].

All the carbon atoms in a crystalline structure are attracted to each other, acquiring an ordered structure. One of the simplest ways to differentiate carbons found in nature from those that are man-made is according to the degree of ordering of their atoms as shown in Figure 12. At the higher order end is diamond and shortly before that is graphite. Accordingly, a carbon will be more ordered the more its formation process has been carried out at a higher temperature and over a longer period of time.

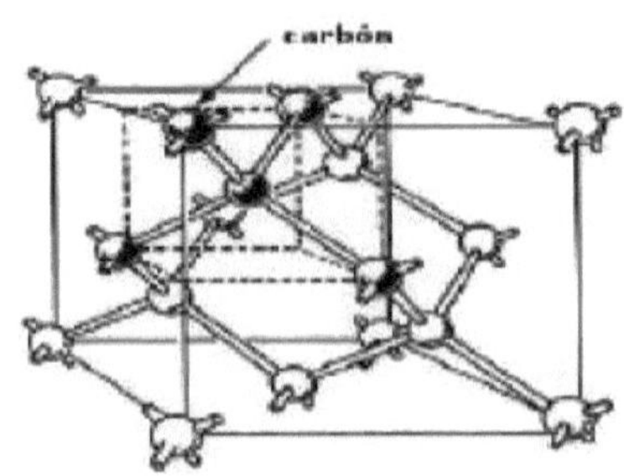

Figure 12. Atomic arrangement of carbon

2.5.2- - History of activated carbon

The use of carbonaceous materials is lost in history, so that it is practically impossible to determine exactly when mankind began to use them. What is certain is that before the use of what we nowadays call activated carbons, i.e. carbons with a highly developed porous structure, charcoal, or simply partially devolatilised or burnt wood, were already used as adsorbents.

The earliest uses of these primitive activated charcoals, generally prepared from charred wood and charcoal, appear to have had medical applications. Thus, a papyrus dating from 1550 BC was found in Thebes Greece describing the use of charcoal as an adsorbent for certain medical practices. In relation to the treatment of water with activated

53

charcoal, it is known that in 450 B.C. on Phoenician ships drinking water was stored in barrels with the wood partially charred on the inside. This practice continued until the 18th century as a means of prolonging the water supply on transoceanic voyages. However, the first documented application of activated charcoal for use with a gas does not occur until 1793, when Dr. D.M. Kehl uses charcoal to mitigate the odours emanating from gangrene. The same doctor also recommended filtering water with charcoal [41].

The first industrial application of activated carbon took place in 1794 in England, when it was used as a bleaching agent in the sugar industry. This application remained a secret for 18 years until the first patent was issued in 1812. In 1854, the first large-scale application of activated charcoal in the presence of a gas took place when the Mayor of London ordered the installation of charcoal filters in the sewer ventilation systems. In 1872, the first masks with activated charcoal filters used in the chemical industry to prevent the inhalation of mercury vapours appeared.

However, the term adsorption was used until 1881 by Kayser to describe how the charcoal trapped gases. Around this time R. Von Ostrejko, considered the inventor of activated carbon, developed several methods to produce such carbons as we know them today, beyond simple carbonisations of organic materials or charcoal. Thus, in 1901 he patented two different methods to produce activated carbon. The first consisted in the carbonisation of lignocellulosic materials with metal chlorides, which became the basis of what is nowadays chemical activation. In the second, he proposed a gentle gasification of previously carbonised materials with water vapour or CO_2, i.e. physical, or more correctly thermal, activation.

The First World War, and the use of chemical agents during this war, brought about the urgent need for the development of activated carbon filters for gas masks. This event was undoubtedly the starting point for the development of the activated carbon industry and a *number of activated carbon filters used not only in the adsorption of toxic gases but also in water purification*. From this moment on, a multitude of activated carbons were developed for the most diverse applications: gas and water purification, medical applications, catalyst support, etc [42].

2.5.3- Types of activated carbons

Carbons can be classified according to particle size into powdered activated carbon (PAC) and granular activated carbon (GAC). PACs have particle sizes smaller than 100 nm, with typical particle sizes between 15 and 25 nm. GACs have an average particle size between 1 and 5 nm. GAC can be divided into two categories: (i) lump or unshaped activated carbons and (ii) shaped or specifically shaped activated carbons, cylinders, discs, etc., as shown in Figure 3. Lump activated carbons are obtained by grinding, sieving and sorting coal briquettes or larger pieces. Shaped coals can be obtained by pelletisation or by extrusion of powdered charcoal mixed with different types of binders [43].

Figure 13. Classification of activated carbon

Within the classification of lump coals there are [44] the following basic types mentioned in Table V

Table V. Classification of lump coals

Mineral coals	Charcoal
Anthracite	Wood (pine, acacia)
Bituminous coal	**Coconut husk**
Lignite	Bagasse (pineapple, olives, etc.)
Peat	Fruit pits (peach, etc.)

Within the classification of lump coals mentioned in Table 5, the main ones can be described:

Lignite: Formed after peat has been compressed. It is the coal with the lowest calorific value, because it was formed in more recent times and contains less carbon (30%) and more water. It is a brown, crumbly substance in which some plant structures can be recognised.

Hard coal: It originates from the compression of lignite. It has an important calorific value, which is why it is used in energy production plants. It is hard and brittle, black in colour. The concentration of carbon is between 75 and 80%.

Anthracite: it comes from the transformation of hard coal. It is the best of the coals, very little polluting and with a high calorific value. It burns with difficulty but gives off a lot of heat and little smoke. It is black, shiny and very hard. It has a concentration of up to 95% carbon.

Charcoal: This type of charcoal is perhaps the first material used by man and its use probably dates back to the very beginning of the use of fire; given that the pieces of charred wood that remain in some bonfires can be considered rudimentary charcoal. Charcoal is used not only for domestic but also for industrial purposes, especially in developing countries.

2.5.3.1- Properties of activated carbons

There are two fundamental characteristics on which the applications of activated carbon are based: high substance removal capacity and low retention selectivity.

The high capacity for substance removal is due to the high internal surface area, porosity and pore size distribution play an important role. In general, micropores provide the high surface area and retention capacity, while mesopores and macropores are necessary to retain large molecules, such as dyes or colloids, and to facilitate access and rapid diffusion of the molecules to the inner surface of the solid.

On the other hand, activated carbon has low specificity to a retention process, it is a "universal" adsorbent. However, due to its apolar nature y por the type of forces involved in the adsorption process, it will preferentially retain apolar and high molecular volume molecules such as hydrocarbons, phenols, dyes, while substances such as nitrogen, oxygen and water are practically not retained by carbon at room temperature. This is why more than 190,000 tonnes per year of the 375,000 tonnes produced are used to remove pollutants from a wide variety of sectors, both in the gaseous phase (hydrogen sulphide, sulphur dioxide, nitrogen oxides, petrol vapours, etc.) and in the liquid phase (drinking water, industrial and waste water, dyeing plants, etc.).

The adsorptive properties of an activated carbon do not depend solely on surface area and porosity. In particular, polar substances are weakly retained on the non-polar surface of the carbon. In these cases, the attachment of heteroatoms, mainly oxygen and hydrogen, to the carbon, forming structures or functional groups such as carboxylic acid, lactose, carbonyl, etc., increases the affinity of the polar substances for the surface of the adsorbent and can give the carbon an acid-basic character.

Finally, when the substance to be removed has an appreciable polarity, low molecular volume and is very dilute in air, retention at ambient temperature by carbon is only effective if it is impregnated with specific reagents or the catalytic properties of carbon are exploited. Thus, after adsorption, chemical reactions take place which transform the toxic products into inert products which are desorbed or retained in the porosity of the charcoal.

2.5.4- Importance of the porous texture of activated carbons

Activated carbons can have a high surface area, in the order of 500 m^2 /gr and even up to 3000 m^2 /gr. The high values of the spedific surface area are largely due to the porosity of the carbonaceous materials, with micropores being the main contributors to the spedific surface area.

In principle, one would think that the greater the specific surface area, the better the adsorption characteristics of the activated carbon, since we should also have a greater number of centres for adsorption. Thus, depending on the size of the adsorbate molecules, it may happen that these are larger than some of the pores and therefore not all of the surface area is accessible to these molecules [42].

On the other hand, both the pore geometry and the adsorbate geometry have to be taken into account. It has also been observed on numerous occasions that certain compounds adsorb very well on a certain activated carbon, while adsorption is much lower on other activated carbons, despite the fact that these have a similar porous texture, pore size distribution and surface area.

This is due to the important fact that: A high specific surface area, with a good pore size distribution that makes it easily accessible to the adsorbate is a necessary, but not sufficient, condition for optimising the preparation of an activated carbon.

According to the IUPAC (International Union of Pure and Applied Chemists), pores can be classified according to their diameter as follows:

micropores : smaller than 2 nm

mesopores : between 2 and 50 nm

macropores: larger than 50 nm Up to 200 to 2000 nm

The micropores are of a suitable size to retain small molecules, which roughly correspond to compounds more volatile than water, such as odours, flavours and many solvents.

Macropores trap large molecules, such as those that are coloured or humic substances - humic and fulvic acids - that are generated when organic matter decomposes. Mesopores are suitable for molecules in between, as can be seen in Figure 14.

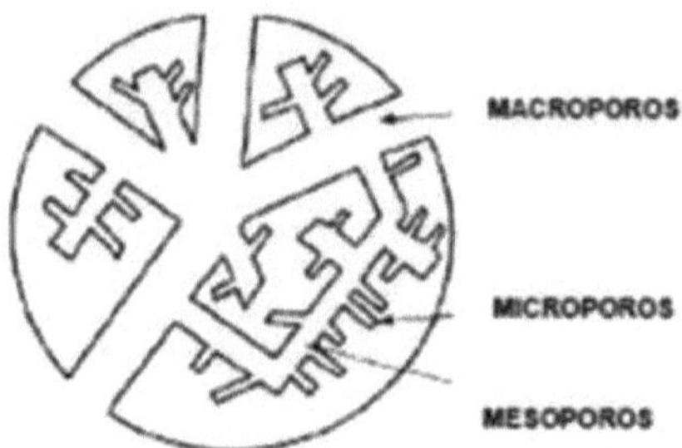

Figure 14. Schematic representation of an activated carbon granules

However, the adsorptive properties of an activated carbon are not only defined by its porous structure, but also by its chemical nature.

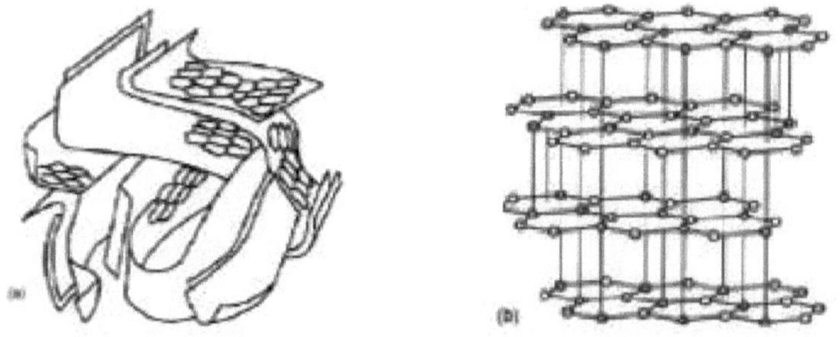

Figure 15.(a) Schematic representation of the microstructure of activated carbon.

(b) Carbon layer accommodation

Activated carbon has in its structure carbon atoms with unsaturated valence as shown in Figure 15, and in addition, functional groups (mainly oxygen and nitrogen) and inorganic components from the ashes, all of which have an important effect on the adsorption processes.

Functional groups are formed during the activation process by interaction between the free radicals on the carbon surface, which causes the carbon surface to become chemically reactive, and is the reason why they affect the adsorptive properties, especially for molecules of a certain polar character [42].

Thus, activated carbon can be considered in principle as hydrophobic, due to its low affinity to water, which is very important in gas adsorption applications in the presence of humidity, or species in aqueous solution; but the presence of functional groups on its surface means that they can react with water, making the surface more hydrophilic. The oxidation of a carbon produces the formation of hydroxyl (OH-), carbonyl (RCHO), carboxyl (RCOOH) groups, etc., which give the carbon an amphoteric character, that is, both acidic and basic at the same time. This influences the adsorption of many molecules [45].

2.5.5- Activation processes

2.5.5.1- Physical process

The process starts with the carbonisation stage, so that dehydration and devolatilisation are achieved in a controlled manner, resulting in a carbonate with a high fixed carbon content and an initial porous structure. During carbonisation the non-carbonaceous elements, such as hydrogen and oxygen, present in the raw material are partly removed by pyrolysis of the material and the carbon atoms are organised into microcrystalline structures known as elementary graphitic crystallites. Between these microcrystals there are free spaces, due to their irregular arrangement.

These spaces or interstices are blocked by amorphous carbon, tars and other residues from the pyrolytic decomposition of the cellulosic material.

As a result, the carbons produced by carbonisation have only a small absorption capacity, which is increased by the activation process.

The activation is carried out in a second stage at temperatures between 800 and 1100°C in the presence of an oxidising agent as activating agent, which can be CO2 and water vapour. The oxidation of the amorphous carbon and the non-uniform gasification of the microcrystals leads, in the first activation phase, to the formation of new pores, i.e. the development of a microporous structure.

It can be seen that there is a strong relationship between the absorption capacity, as a function of the development of the porous structure, and the gasification of the carbonaceous material. Therefore the term "burn off" is used as a measure of the degree of activation and indicates the weight percent decrease of the material during activation.

According to Dubinin 2005, for combustion losses lower than 50% microporous activated carbons are obtained, if the combustion loss is higher than 75% a macroporous carbon is obtained, if the combustion loss is between 50 and 75% the product obtained is a mixture of micro and macroporous structure.

The physical process of activation consists of carbonising the raw material to red heat to expel the hydrocarbons, but without sufficient air to reach combustion, thus obtaining a primary carbon. This is then exposed to an oxidising agent, usually water vapour. The reaction is endothermic, so it is necessary to generate a constant temperature (800-1000 °C).

The physical process is also known as thermal activation although the activation occurs due to a chemical reaction of the activating agent (an oxidant such as air, water vapour, CO2, etc.).

Sometimes certain pre-treatments such as grinding and sieving are necessary to obtain a suitable size of the precursor. If the precursor is a coking carbon, then an oxidation step is necessary to remove the coking properties. In other cases, the starting material is ground to a fine powder, then compacted with a binder to form briquettes, and then ground again to the desired size. In this way, a better diffusion of the activating agent is achieved and thus a better porosity of the resulting activated carbon.

Another stage prior to activation itself is carbonisation, in which the precursor is subjected to high temperatures (of the order of 800°C) in the absence of air, in order to remove volatiles and leave a carbonaceous residue which will be subjected to activation. During devolatilisation, the release of gases and vapours from the precursor produces an "incipient" porosity in the carbonate, which develops further during the activation stage.

Activation itself may be a process totally independent of carbonisation or may be carried out after carbonisation. It consists of reacting the activating agent with the carbon atoms of the carbonised material being activated; in such a way that a "selective burning" is produced which progressively perforates the carbonised material, generating pores and increasing the porosity until it is transformed into an activated carbon. The activating agents that are usually used are: Oxygen (rarely on an industrial scale), air, water vapour (the most commonly used) and CO_2. These agents give rise to the following chemical reactions where carbon atoms are eliminated, thus producing porosity:

$$C + O_2 \rightarrow CO_2 \tag{1}$$

$$C + H_2O \rightarrow H_2 + CO \tag{2}$$

$$2C + O_2 \rightarrow CO_2 \tag{3}$$

$$C + CO_2 \rightarrow 2C \tag{4}$$

2.5.5.2- **Chemical process**

This process is carried out in a single stage, by heating a mixture of the activating agent and the starting material in an inert atmosphere. The most commonly used substances are: phosphoric acid (H3PO4), zinc chloride (ZnCh), sulphuric acid (H2SO4), although potassium sulphides and thiocyanates, calcium and magnesium chlorides, alkali metal hydroxides, among other substances, have also been used, always depending on the original raw material to be used and the greater or lesser volume of pores of one type or another to be obtained.

Among the raw materials of vegetable origin, mainly wood sawdust and phosphoric acid (H3PO4) are used as activating agent. Considering that the sawdust is waste and the activating agent can be recovered, it makes the process commercially viable, although other raw materials such as mineral coals are also used. The process involves mixing the original raw material with the activating agent (dehydrating agent), forming a paste which is then dried and carbonised in a kiln, at a temperature between 200 and 650 °C, with dehydration occurring with the end result of creating a porous structure and an enlarged surface area. The fundamental parameters controlling the chemical activation process and the product to be obtained are: the impregnation ratio, the activation temperature and the residence time. The dependence of the carbon structure on these variables can be followed by changes in the shapes of the adsorption isotherms.

Chemical activation is based on dehydration by means of chemical substances at a medium temperature (400 - 600°C). This depends on the chemical substance to be used to activate the carbon.

In this type of activation the precursor is reacted with a chemical activating agent. In this case, the activation usually takes place in a single step at temperatures ranging from 450 to 900°C. However, a subsequent washing step of the activated carbon is necessary to remove any traces of the activating agent.

There are numerous compounds that can be used as activating agents, but the most widely used industrially are zinc chloride ($ZnCl_2$), phosphoric acid (H_3PO_4) and potassium hydroxide (KOH).

Chemical activation with ZnCh was the most widely used method until 1970, especially for the activation of wood residues. Its use, however, has been greatly restricted nowadays, due to the environmental problems associated with the use of $ZnCl_2$. However, some countries, such as China, still use this method to produce activated carbon.

Chemical activation with H_3PO_4 has practically displaced ZnCh and the precursors in this type of activation are mostly, as in the case of $ZnCl_2$, forest residues (wood, coconut husks, olive pits, etc.).

Activation with H_3PO_4 involves the following steps:

I. Grinding and sorting of the starting material,

II. Mix the precursor with H_3PO_4 (recycled and fresh),

III. Heat treatment in an inert atmosphere between 100 and 200°C, maintaining the temperature at approximately 1h, followed by a further heat treatment up to 400500°C, maintaining this temperature at approximately 1h,

IV. Washing, drying and sorting of activated carbon, and recycling of H3PO4. The most commonly used H3PO4:precursor ratio is usually 1:5 (although different ratios result in carbons with different properties) and the activated carbon yield is usually 50%.

The chemical activation with KOH was developed during the 1970s to produce so-called "superactivated coals", with specific surface areas in the order of 3000 m^2 /gr. Unlike the other two activating agents, the preferred precursors for activation with KOH are those with low volatiles and high carbon content, such as high-grade mineral coals, carbonised coals, petroleum coke, etc. In this activation the KOH is mixed with the precursor, in an aqueous suspension or by simple physical mixing, in proportions between 2:1 and 4:1. When the impregnation takes place in an aqueous medium, the activation is carried out in two consecutive heat treatments in an inert atmosphere. The first one at low temperatures, but above 200°C (used only to evaporate the water and disperse the KOH) and the second one between 700 and 900 °C [40].

2.5.6- Adsorption with activated carbon

2.5.6.1- Description of adsorption

In general terms, the adsorption process consists of the capture of soluble substances present at the interface of a solution, where the interface may be between a liquid and a gas, a solid or between two different liquids.

The use of the term sorption is due to the difficulty of differentiating between physical and chemical adsorption, and is used to describe the mechanism by which organic matter binds to GAC. Equilibrium is reached when the rates of sorption and desorption are equalised, at which point the adsorption capacity of the carbon is exhausted. The theoretical adsorption capacity of a given pollutant by activated carbon can be determined by calculating its adsorption isotherm.

The amount of adsorbate that can be retained by an adsorbent is a function of the characteristics and concentration of the adsorbate and the temperature. In general, the amount of adsorbed matter is determined as a function of concentration at constant temperature, and the resulting function is known as the adsorption isotherm [46].

The molecules in the gas or liquid phase will be physically bound to a surface, in this case the surface is activated carbon. The adsorption process occurs in three steps:

- Macrotransport: Movement of the organic material through the macropore system of the activated carbon.

- Microtransport: Movement of the organic material through the micropore system of the activated carbon.

- Adsorption: Physical adhesion of the organic material to the surface of the active carbon in the mesopores and micropores of the active carbon.

The level of adsorption activity depends on the concentration of the substance in the water, the temperature and the polarity of the substance. A polar or water-soluble substance cannot be removed or is poorly removed by activated carbon.

2.6- CYANIDATION AND ITS EFFLUENTS

2.6.1- **Remediation of cyanide in solution**

Cyanidation effluent treatments are divided into cyanide destruction processes, alkaline chlorination, sulphur dioxide/air, hydrogen peroxide, biodegradation, cyanisorb process, and other methods:

a) Cyanide destruction processes

Alkaline chlorination: This method, the oldest in the treatment of cyanidation effluents, consists in the oxidation and destruction of free cyanide and some complex cyanides by means of hypochlorite or chlorine under alkaline conditions pH = 10.5-11.5. In practice, chlorine consumption is very high (oxidation of cyanide to cyanate requires 2.75 parts by weight of chlorine to one part cyanide) and does not remove complex iron and cobalt metal cyanides. Another disadvantage is the production of highly toxic chlorination residues that must be removed before effluents are discharged to the final site [47-48].

SO2/Air process: Known in two versions, INCO process the most used and Noranda process, it is characterised by the oxidation of cyanide with SO2/air, at pH between 8 and 10, between 5 and 60 °C, and the use of copper in solution as a catalyst. The residual metals released from the single cyanides are generally precipitated as hydroxides. The method allows the oxidation of a part of the thiocyanate; however, its total removal requires long retention times and addition of more reagent. This method requires high consumption of SO2 generally produced *in-situ* and supplementary reagents such as sodium sulphite and sodium meta-bisulphite [49-51].

Hydrogen Peroxide: In this process (known as Degusta and Kastone), cyanide and cyanide complexes are oxidised to cyanate (with precipitation of the metal ions) by using H2O2 and copper as a catalyst.

With this method 10 to 20 % of the cyanate produced from the oxidation of cyanide is converted to ammonium and only 10 to 15 % of the thiocyanate is oxidised. In practice, 200 to 450 % in excess hydrogen peroxide is used. Effluents from this process require further treatment in cases where they contain significant levels of thiocyanate and ammonia, as well as copper [49, 51].

b) Biodegradation

Biological processes, whose interest has increased considerably in the last decades, are generally applied to effluents whose concentration of weak cyanide with respect to free cyanide is much lower (< 50 mg/L). The process is carried out in 2 bacteriological oxidation stages: the breakdown of cyanide and thiocyanate by oxidation, and the adsorption-precipitation of metals in the bio-layer and nitrification of ammonia. Metals and suspended solids incorporated in the bio-layer and metal precipitate sludge are removed by clarification of the solution. In general, the biodegradation process is carried out at temperatures above 10 °C and the treatment time varies from 20 to 60 days, depending on the growth rate of the microorganisms and the accumulation of metals in the bio-layer [49, 52, 51].

c) Cyanisorb process

The recovery of cyanide through the stages of acidification, volatilisation and reabsorption was first known as the Merrill-Crower process. Later, some modifications gave rise to the AVR process: acidification-volatilisation-reneutralisation, or Cyanisorb.

In general, the process consists of acidifying the cyanide in a clarified solution by lowering the pH from 8.5 to less than 2, volatilising the HCN by aeration in tanks or closed vessels requiring large volumes of air and pressure of less than 0.5 lb/in2 , and reabsorbing the HCN gas in a caustic solution at a pH between 10.5 and 11.5. The recovered cyanide is reused in the cyanidation process. Although this process allows for cyanide recovery efficiency above 95%, its main disadvantage is the imminent danger related to leakage of the lethally toxic HCN gas, so safety measures must be very strict [53].

d) Other methods

There are other types of cyanide effluent treatment processes used on a smaller scale or under development. Among these processes are ion exchange used to remove metal cyanide complexes, precipitation of free cyanide with ferrous ion, adsorption of free cyanide and metal cyanides with ferrous sulphide; adsorption on activated carbon, flotation-precipitation and ionic, electrochemical methods, chemical oxidation of cyanide with ozone or its recovery from thiocyanate [54].

The treatment of effluents from the cyanidation process for the extraction of gold and silver is one of the main problems facing the precious metals industry. These effluents contain varying amounts of free cyanide, weak and strong cyanide complexes and thiocyanate. Different methods exist to remove cyanide from these effluents, but most of the time the reagent consumption increases the operating costs to levels too high for a cost-effective process. Some of the available methods present the problem of producing by-products that are also toxic.

3.1- Materials, reagents and equipment used in the experiments

<u>Material</u>
Burettes 25, 50 ml
Funnels
Filters
Volumetric volumetric volumetric flask of 10, 25, 50, 1000ml.
Suction knob
Pipettes 10, 15mL
Test tubes 10, 25, 50, 100, 250 ml.
Universal Supports
Thermometer
Beakers 10, 50, 100, 100, 250, 500, 1000ml.
Magnetic stirrers
Vials
Stopwatch
Droppers
Amber coloured glass containers50, 100, 1000 ml.
Plastic containers of 25, 50, 100, 500, 1000 ml.
Erlenmeyer flask 50, 100, 250 ml.

<u>Reagents</u>
Sodium cyanide (NaCN), Jalmex purity 95%.
5-[4-(dimethylamino-benzylidine] rhodanine ($C_{12}H_{12}N_2OS_2$), Aldrich 97%, -5-[4-(dimethylamino-benzylidine] rhodanine ($C_{12}H_{12}N_2OS_2$), Aldrich 97%.
Silver nitrate ($AgNO_3$), Faga Lab
Deionised water, Fermont

<u>Equipment</u>
Analphabetical balance, explorer ohaus
pH meter model 720 A Orion
Rotap and sieves, Gilson Company
Magnetic stirring grill, thermo scientific

<u>Materials</u>

Coconut shell activated carbon

Barite

Fluorite

Molybdenite

Calcite

3.2 Grinding of coconut shell activated carbon

Firstly, coconut husk samples were taken, with a total weight of 10g each, and ground for 2 hours at a speed of 20 rpm, to observe the best grinding time, with 0.5 hours of rest for each hour of work.

The components and steel balls were loaded into a stainless steel vial, grinding was carried out in a Restch mill to produce powder from the different samples to be treated, the total weight of the sample was 10g.

In the mechanical milling process, it was necessary to consider 0.5 hours of rest for each hour of work, which was due to the heating of the mill motor.

Once a large amount of powder was obtained from the mechanical milling, it was stored in glass vials with lids to avoid any contamination.

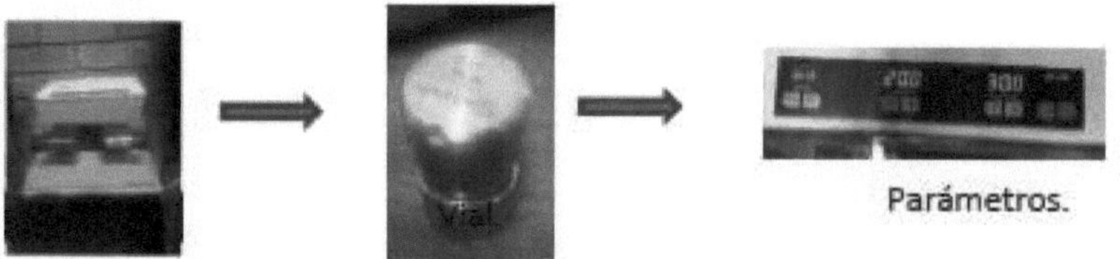

3.3 Determination of free cyanide

Free cyanide is determined by silver nitrate titration, according to the APHA-AWWA 4500-CN D method. The basic principle of this technique is the determination of the cyanide ion by titration with silver nitrate ($AgNO3$) to form a soluble silver cyanide complex ($Ag\ (CN)^{2-}$). When all the cyanide is in the form of such a complex and there is a small excess of silver, this is detected by an indicator, changing the colour of the solution. The reaction that occurs is as follows:

$$2CN^- + AgNO_3 \rightarrow Ag(CN)_2^- + NO_3^-$$

Table VI Reagents required for free cyanide analysis.

Substance	Grade	Manufacturer
Silver nitrate (AgNOs)	99.3 %	Aldrich
5-[4-(dimethylamino)-benzylidene]rhodanine	97%	Aldrich

To prepare the 0.0038 M AgNO3 solution, weigh 0.6522g of AgNO3 and make up to 1l with deionised water. As for the indicator solution C12H12N2OS2; weigh .02 g and make up to 100 ml with Acetone C3H60 (manufacturer ANALYTYKA).

The procedure for the determination of free cyanide in solution was as follows:

1. 10 ml of the test sample was taken so that the analysis could be carried out within the limits set by the calibration curve.

2. The solution was previously verified to have a pH greater than 11.

3. The aKquota was placed in a 50 ml beaker and 3 drops of the C12H12N2OS2 indicator solution.

4. The sample was titrated with 0.0038 M AgNO3 solution.

5. When a change of colour occurs in the sample, the titration was stopped and the volume of AgNO3 consumed was read.

6. The corresponding calculations have been made, according to the following relationship:

1 ml of added AgNO3 = 20 mg/L of CN⁻

7.   Corrections were made for the dilution factor used.

3.4 *Preparation for cyanidation.*

a.- The materials from the milling and the thermo-chemical treatment in the case of coconut shell carbon *are* prepared to be put in contact with cyanide.

b.- Tridistilled water was prepared, adjusting the pH to 11 with sodium hydroxide.

c.- NaCN solutions were prepared. In a 1 L volumetric flask and add 400 ppm of CN.

d.- subsequently, 250 ml of sodium cyanide are quantified and placed in a beaker with a magnet, and with the help of a stopwatch, the experiment is started.

e.- at the end, the remaining carbon sediment is filtered and stored for later use.

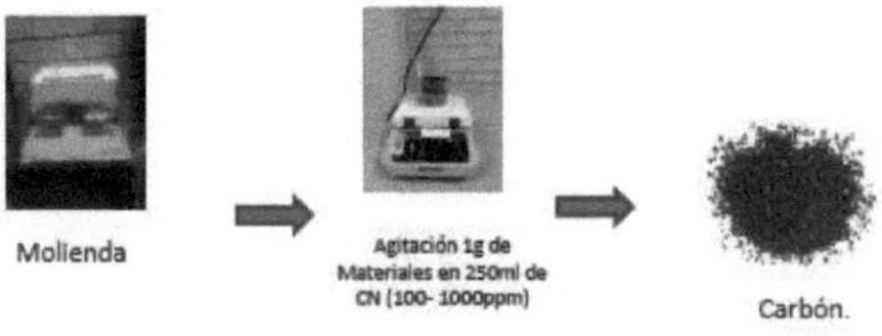

4.1- CYANIDE ADSORPTION KINETICS

Cyanide solutions 200 ppm were put in contact with different materials coming from mechanical milling, which resulted in Figure 16, describing that the material with the best adsorption kinetics was the mineral molybdenite, this comes from a depletion flotation where copper is extracted, molybdenite MoS2-Cu has a + charge and therefore there is a good adsorption of cyanide, with a maximum adsorption of 53% compared to coconut shell carbon which is 50%, fluorspar, barite and calcite with a maximum percentage of 32, 30.5 y 30.5.

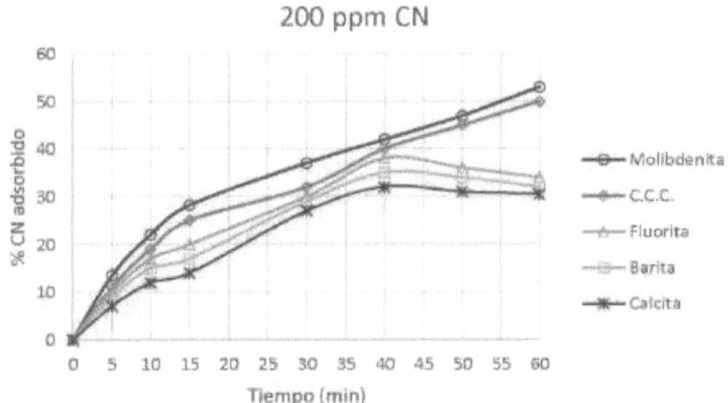

Figura 16.   Cyanide adsorption on different materials at 200 ppm cyanide

Figure 17 is a concentration of 400 ppm of cyanide that were put in contact with the same materials but the maximum adsorption peak was again given by molybdenite 56% presenting the same adsorption phenomenon followed by coconut shell carbon with 54, compared to the previous figure fluorite, calcite and barite had a better adsorption 45, 38, 34 % may be mainly due to the trace metals they bring.

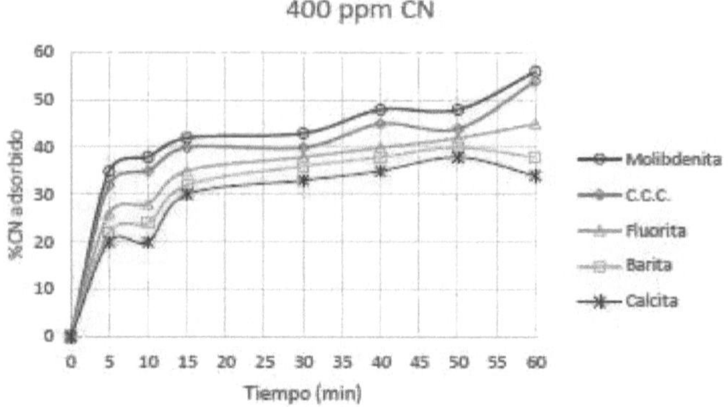

Figura 17. Adsorption of cyanide in different materials at 400 ppm cyanide

With respect to the concentration of 600 ppm CN, there was no great improvement compared to Figure 16, since both molybdenite and coconut shell carbon have values of 56 and 54, possibly due to a supersaturation of the pores of both materials. This phenomenon of destabilisation has also been reported by other researchers such as Reinoso and Castilla.

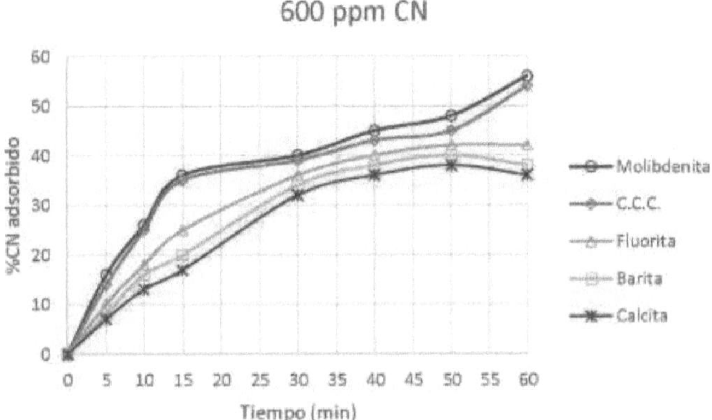

Figura 18. Cyanide adsorption on different materials at 600 ppm cyanide

With respect to the 800 ppm concentration of cyanide, it also presents the phenomenon of ion exchange, the percentage of adsorption is 58 and 56 for coconut shell carbon, in comparison with the other figures, fluorspar, barite and calcite adsorbed 37, 34, 34 as shown in Figure 18.

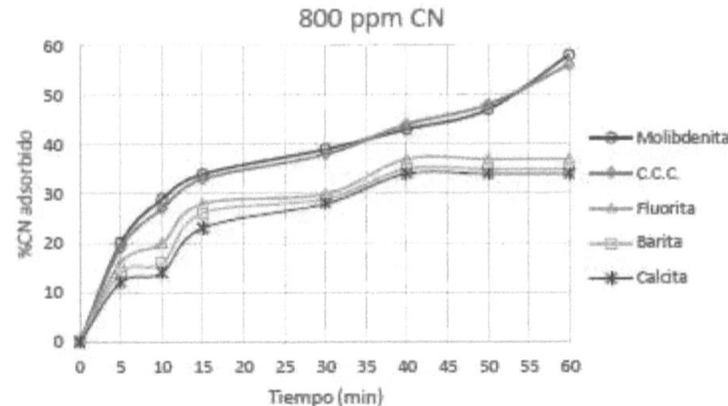

Figura 19. Adsorption of cyanide on different materials at 800 ppm of cyanide Once the maximum adsorption peaks were evaluated at concentrations of 200, 400, 600, 800 ppm of cyanide, a decision was made to work with this concentration but with coconut shell activated carbon with heat treatment. The decision was taken to work with 400

ppm in order not to use so much cyanide, as the adsorption percentages were almost similar in the three cases. The coconut shell carbon and the best cyanide adsorbent of the previous stages were put in contact for 24 hours, and the result was that after 1 hour of adsorption the adsorption percentages were very similar, but after 2 hours of adsorption the modified coconut shell carbon presented better adsorption, this could be due to the functional groups increased in the thermal treatment, it is known that coconut shell carbon is of a basic nature and the quinones and carbonyl functional groups predominate.

Once the 24 hours had elapsed, the experiment was stopped because there was not so much liquid in solution and the adsorption did not have a big difference, giving a maximum adsorption percentage of 78% as shown in Figure 19.

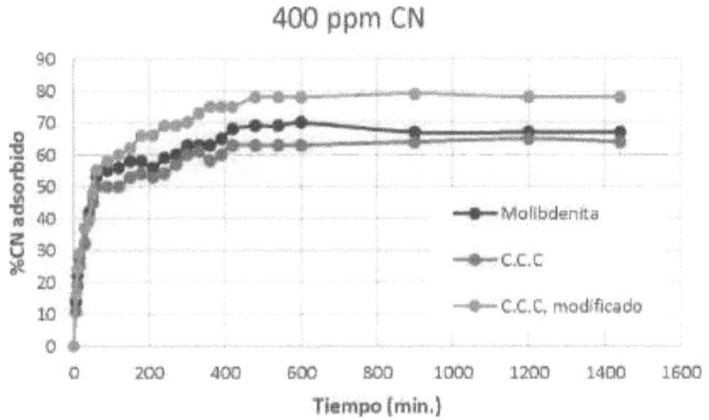

Figura 20. Adsorption of cyanide at 400 ppm cyanide with molybdenite, coconut shell charcoal
and modified
coconut shell charcoal

4.2 Characterisation of the adsorbent by Fourier Transform Infrared Spectroscopy

Figure 21 shows the IR spectrum obtained from the experiment performed at 400ppm cyanide. The broadening of the spectrum shows two bands between 1020 and 934 cm^{-1} , with maxima at 1044.5 and 970.5 cm^{-1} . These bands correspond to the symmetric and antisymmetric vibrations respectively of the O-Mo-O stretching, and a slight shift to lower values of the transmittance percentage may be due to the adsorption of CN- on the molybdenite.

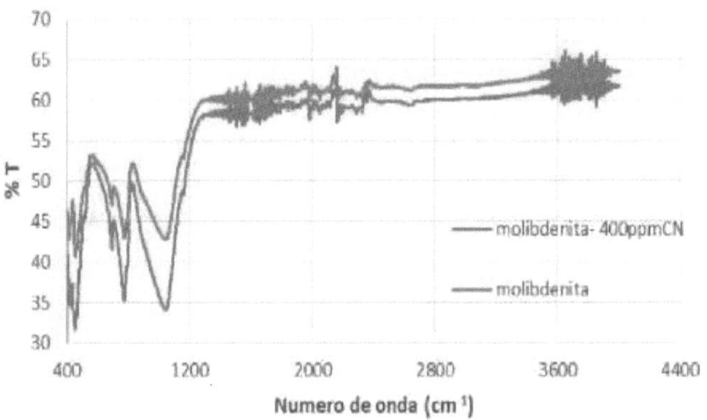

Figura 21. Infrared spectrum of molybdenite before and after cyanidation process
cyanidation process

FTIR analysis of coconut shell carbon in Figure 22 reveals that it has a broad band due to the formation of hydrogen bridges, with a peak around 3440 cm^{-1} and is related to the tension-type vibration of the O-H hydroxyl groups, overlapping with N-H bonds: alcohols-phenols, carboxyl and amino [55] which is corroborated by the formation of signals at 1500 and 1532 cm^{-1} for phenols and alcohols.

and alcohols. The bands at 1555 and 1577-1240 cm^{-1} are attributed to asymmetric and symmetric bending of the CH3 group.

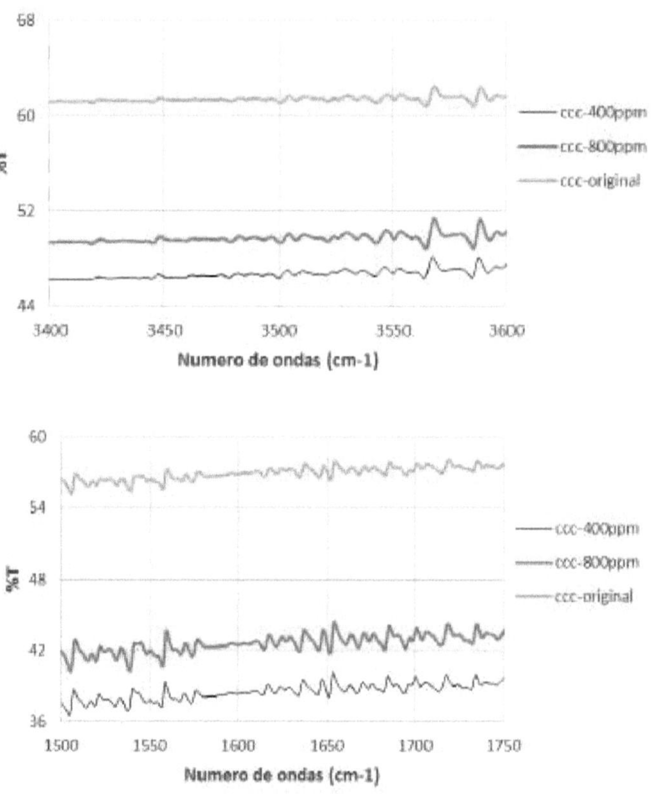

Figure 22 Carbon spectra of coconut shell without modification and in contact with cyanide a) stretching 3400-3600 b) stretching 100 -1750

Figure 23 shows the FTIR spectrum of the C.C.C.; it can be seen that the precursor has undergone important modifications in its structure, suggesting the complete carbonisation of the precursor when subjected to the activation process. Volatile and organic compounds have been decomposed or removed, resulting in an efficient modification of the carbon surface [56, 57]. The broad band of the precursor between 3600 and 3050 cm^{-1} , as well as the sharp bands at 2922, 2853, 1743 and 1032 cm^{-1} have practically disappeared. Only a small peak of weak intensity has remained around 2980 and 2900 cm^{-1} products of aliphatic C-H stretching in an aromatic methoxyl group [55] and the peaks observed around 1560 and 1150-1060 cm^{-1} are due to C=C and C-C stress-type vibrations respectively. The band peaking at 1560 cm^{-1} is superimposed with the bending vibration of the N-H group (amines and amides). The peaks at 1150-1060 cm^{-1} are related to the superposition of C-O stress vibrations of ethers, esters [57] with shoulder at 1190 cm^{-1} ; they are characteristic signals of phosphorcarbonaceous compounds present in phosphoric acid activated carbons [55].

The signal at 870 cm-1 is related to bending vibrations out of the C=C-H plane [57], however, it has undergone a slight shift to lower values of the transmittance percentage. This may be due to the adsorption of CN- on the active carbon.

The formation of two sharp bands at 2922 and 2853 cm-1 is attributed to the C-H stress vibration (alkanes) [56]. The peaks at 1743 and 1647 cm^{-1} are assigned to C=O stress vibrations in carboxylic acids, an^drides and lactones [59] and of the acetyl group in hemicellulose [58]; the band at 1240 cm-1 is also associated with C-O stretching of the aryl group in lignin [58] .The signals between 666 and 871 cm^{-1} correspond to out-of-plane bending vibrations [57] and of the N-H amino and C-H aromatic groups [55].

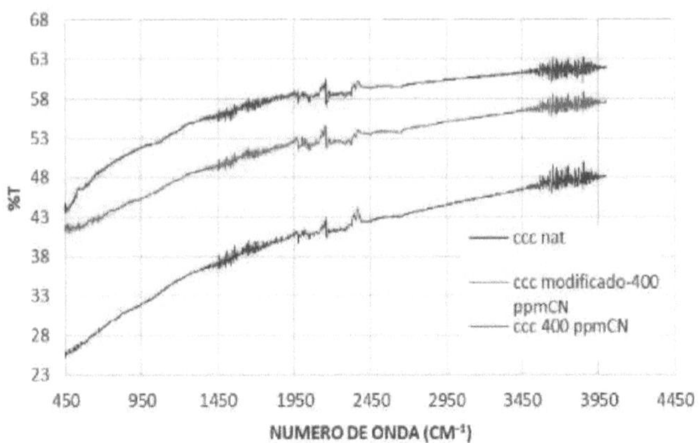

Figure 23 Carbon spectra of modified coconut husk carbon in contact with 400 ppm cyanide

4.3.- Characterisation of coconut shell activated carbon

The American Society for Testing and Materials Standards (ASTM) establishes standard tests necessary to establish the quality of the charcoal obtained, including: percentage of moisture, volatile matter, ash, fixed carbon, and resistance to abrasion. For the characterisation of the porous structure of the activated carbon, some simple techniques are usually used, such as the iodine index, since the carbons are products with a very high capacity to retain contaminants from various fluids. Table VII shows the characterisation of carbon before chemical treatment.

Table I Analysis of the coconut shell charcoal

Technical data	Coconut shell carbon	Standard
Number of iodine	878 mg L/g	ASTM D-4607
Hardness	78 dimensionless	ASTM D-3802
Total ash	3 %	ASTM D-2866

	.35 g/cm3	ASTM D-2854
Bulk density		
Total humidity	2 % max	ASTM D-2867

The textural properties of the carbon (specific area, pore volume and average pore diameter) were determined by means of N2 physisorption equipment and resulted in Table VIII.

Table IIIII Specific surface area of charcoal

Charcoal	S BET (m^2 /g)
Coconut shell carbon	842
Modified coconut shell carbon	960

The surface area of the carbons used in the investigation is in the Upic range of activated carbons from 500 to 1500 m^2 /g [41] for this reason it is considered to have the potential to be used as an adsorbent or as a precursor for modified adsorbents and applied to the removal of dissolved organic or inorganic pollutants in aqueous media.

4.3.1 X-ray photoelectron spectroscopy.

X-ray photoelectron spectroscopy is a technique for identifying functional groups on the surface of carbon, among other uses. The equipment used in the research was a Thermo Scientific, K-Alpha surface analysis, with a database called Thermoavantage and with a beam diameter of 400 pm and a depth of 10 nm.

Figure 24 shows a full scan of the modified coconut shell carbon sample in contact with 400 ppm cyanide, showing the characteristic peaks of O1S (525-545) and CIS (280-295).

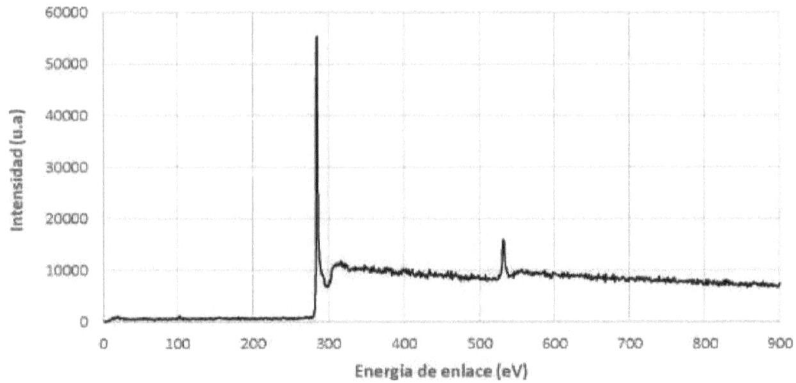

Figure 24. Full scan of coconut shell charcoal with characteristic peaks of O1S and
characteristic peaks of O1S and CIS

A deconvolution of the figure for the CIS was performed, resulting in the following Figure 25, showing the peaks of the different functional groups in which the cyanide is bonded to the carbon. The peak present at 288.7 eV, is the characteristic peak of the carbon, next to this peak there are two other small peaks, which are different functional groups of the carbon, which are interacting with the cyanide:

(1)at 288.7 eV we have (-C=N, -N=C-O-, COOC-).

(2) at 289.3 eV (-COOH) carboxylic acid.

(3) at 291.0 eV we have carbonyl quinone (C=O).

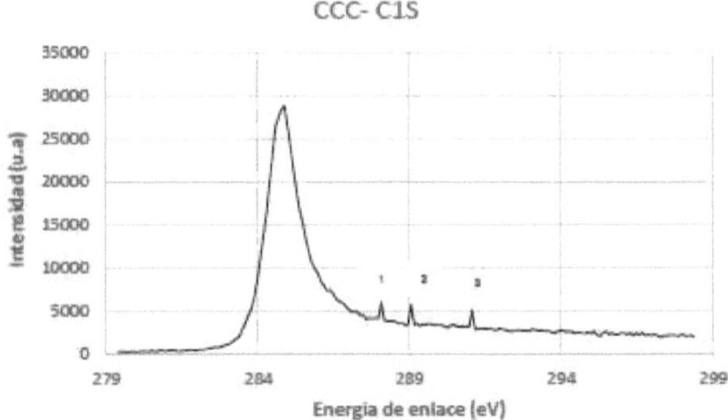

Figure 25 Spectra of coconut shell carbon sample C1S, 1) characteristic carbon peak, 2) carboxylic acid 3) carbonyl quinone], using 400 ppm cyanide.

Few investigations use this technique to identify the functional groups causing the adsorption, because it is an economically expensive technique and most of the investigations are declining to use Fourier transform infrared spectroscopy because it is more economical.

In the literature published in 2000, I used X-ray photoelectron spectroscopy to quantify the functional groups of charcoal in the adsorption of peroxide, showing that the adsorbing groups are the carbonyl groups whose binding energy is 286.1 eV as shown in Figure 26 and in our research were responsible for the adsorption of cyanide.

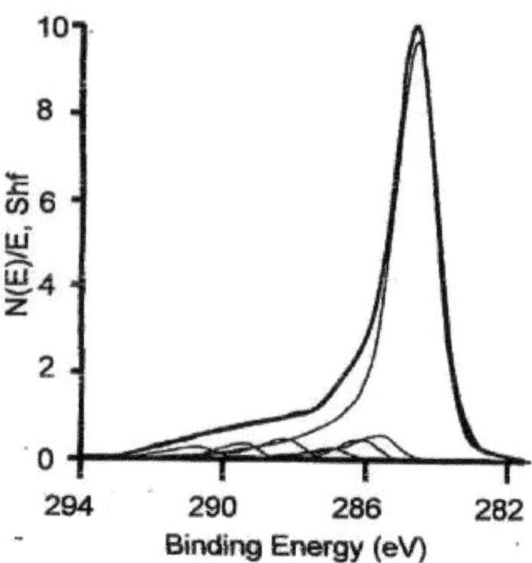

Figure 26 XPS of the BV29H carbon sample (Moreno Castilla et al., 2000).

In the literature published in 2002, they contacted a mineral carbon with ozone and found that the anchor groups responsible for adsorption were found at wavelengths 284.6, 286, 287, 288.6, 291, as shown in Figure 27, and that these groups were also found in the activated carbon used in the research.

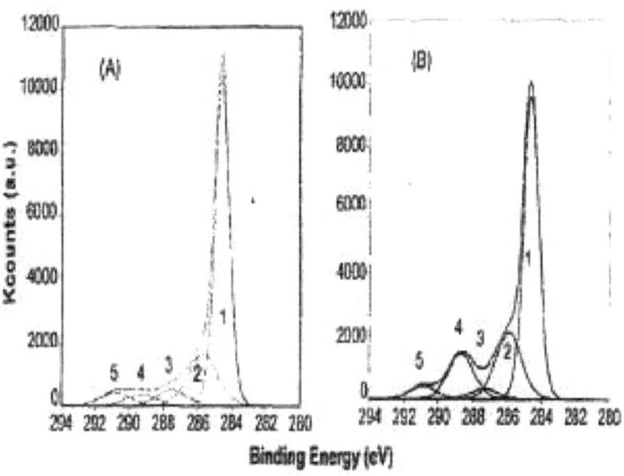

Figure 27 XPS of activated carbon samples in contact with ozone (Valdez, 2002).

In the literature published in 2012, they modified the surface of a coconut shell carbon with ammonia, so that the carbon had more macropores y to adsorb more thiocyanate, in these they found some characteristic binding energies of adsorption, which shows that the binding energies of the carbon indicated that carbonyl groups were introduced to the surface at low oxidation potentials y the concentration of alcohol y ether groups increased at high oxidation potentials 284.5 y 286.1 eV como is shown in FIGURE 28, these groups were also observed in our investigation.

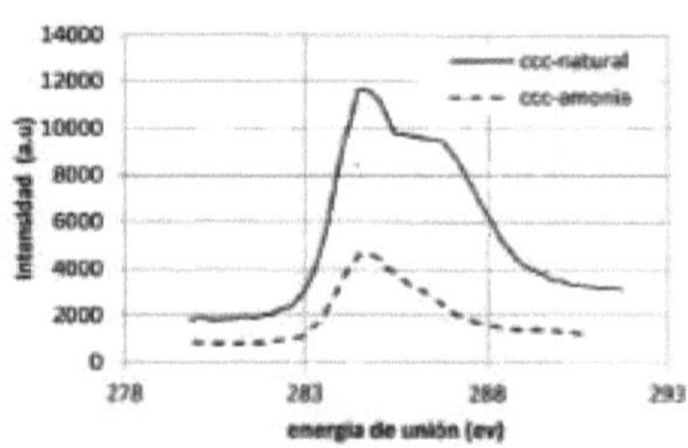

Figura 24.- Espectro de carbón cáscara de coco modificado con amonia.

Tabla 10.-El C_{1s}, O_{1s} y N_{1s} de los espectros XPS para la modificación del ccc-amonia-apelsa

C_{1s}		O_{1s}		N_{1s}	
Ev	Asignación	Ev	Asignación	eV	Asignación
284.5 ± 0.1	-C=C-, No funcionalizado SP^2C	532.3 ± 0.3	C=O (ester, amidas)	399.4 ± 0.3	-N-H , Nitrogeno pirrolico, piridina nitrógeno
286.1 ± 0.1	-C-OH, C-O-CE, C-O-R	533.5 ± 0.2	C-O-CE (ester, oxígeno)	400.2 ± 0.1	-O-C=N, pirrolico piridina
		534.3 ± 0.2	COH, COOH, N-O-N	401.3 ± 0.1	N, cuaternario

Figure 24.- Ammonia-modified coconut shell carbon spectra.

Table 10.-The $C_{|s}$, $O_{|s}$ y $N_{|s}$ of the XPS spectra for the modification of the ccc-ammonia-apelsa

Cis		O is		N_{IS}	
Ev	Allocation	Ev	Allocation	eV	Allocation
284.5 ±0.1	-C=C-, Not functional SP^2 C	532.3 ± 0.3	C=O (ester, amides)	399.4 ±0.3	-N-H , pyrrolic nitrogen, pyridine nitrogen, pyridine nitrogen
286.1 ±0.1	-C-OH, C-O-CE, C-O-R	533.5 ± 0.2	C-O-CH (ester,	400.2 ±0.1	-O-C=N, pyrroiic

			oxygen)		pyridine
		534.3 ±0.2	COH. COOH, N-O-N	401.3 ±0.1	N, quaternary

Figure 28 XPS of coconut shell carbon (castilla, 2012).

4.3.2 Determination of zero load points.

This method is based on the estimation of the zero charge point (ZCP) of the carbon surface from the pH measurement of a suspension in which the amount of suspended solids is progressively increased.

Table IX gives the P.C.C. of the coals used in the experimentation.

Table IX Zero loading points of the coals used in the investigation.

Carbón	P.C.C
C.C.C	5.057
C.C.C MODIFICADO	6.386

There are investigations where mineral coals are used having a higher charge point than the cyanide solution (pH 11), so the attraction that takes place is that the sodium that accompanies the solution is attracted to the surface, on the other hand in our research we tested with vegetable coals and we found that the one that is attracted electrostatically is the cyanide of the solution, this because the pH of the solution is 11 and the charge points are lower, negative charges are attracted to the solution.

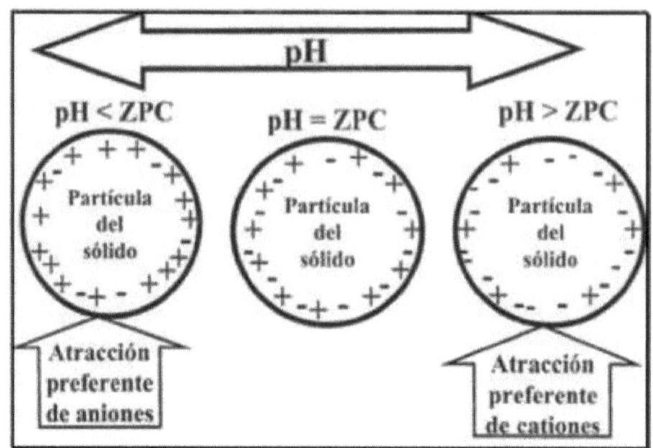

Figure 29 Scheme of representation of the C.C.P.

CONCLUSIONS

The cyanide adsorption kinetics with the different materials, the best one was molybdenite because of the charge in which it works and also the phenomenon that predominated in its adsorption was an interaction of charges, and with the modified activated carbon, a higher percentage of 78 percent could be noted due to the increase of functional groups of the coconut shell carbon.

With respect to the infrared spectroscopy with Fourier transforms, the presence of the functional groups of the activated carbon can be noticed, which were corroborated in XPS, and also the molybdenite shifts could be noticed, this was due to the cyanide adsorption.

With respect to the X-ray photoelectron spectroscopy, they showed a greater presence of the cyano group in most of them, in the characterised samples we also noticed the presence of the carboxylic and carbonyl functional groups, these groups according to the literature are the ones in charge of the attraction of cyanide ion to the carbon surface.

In the determination of the zero charge points of the charcoal, it was found that charcoals with charge points lower than the pH of the solution attract the cyanide ion in the solution more quickly.

BARITA

[1] Lopez, E. (1993). Geolog^a General y de Mexico. Distrito Federal, Mexico: Editorial Trillas.

[2] Dana H. (1959) *Manual de Minera^a (2ª* ed.) Madrid: Reverte

[3] Cornelius, K., Cornelius, S.H. (1997) *Manual of Mineralogy*.

[4] Kelly Errol G. (1990) *Introduction to mineral processing.* Mexico: Limusa

[5] Servicio Geologico Mexicano, SGM (2016), *Anuario Estadistico de la Mineria Mexicana, 2015*, Mexico: SGM.

CALCITE

[6] 2001-2005 Mineral Data Publishing, version 1

[7] http://geology.com/minerals/calcite.shtml

[8] https://mineriaenlinea.com/

[9] https://www.ecured.cu/Calcita

FLUORITE

[10] Real Academia Espanola and Asociacion de Academias de la Lengua Espanola (2014). "fluorina". *Diccionario de la lengua espanola (*23rd edition). Madrid: Espasa. ISBN 978-84-670-4189-7.

[11] Jennie Harding, (2007), *Crystals*, Ed. Press Limited, p. 106.

[12] Hans-Rudolf Wenk, Andrei, (2003), *Minerals: Their Constitution and Origin*, first edition, Cambridge University Press, p. 340.

[13] Fisher, J., Jarnot, M., Neumeier, G., Pasto, A., Staebler, G. and Wilson, T., ed. *Fluorite. The Collector's Choice*. Lithographie, LCC. Connecticut. ISBN 0971537194.

[14] Staebler, G., Deville, J., Verbeek, E., Richards, P. y Cesbron, F. (2006). "Fluorite: from ancient treasures to modern lab and collections". In

Jesse Fisher, ed. *Fluorite. The Collector Choice*. Lithographie, LLC. p. 4-12.

[15] Valeur, B. and Berberan-Santos, M.N. (2011). "A brief history of fluorescence and phosphorescence before the emergence of quantum theory". *Journal of Chemical Education, 88, 731-738*.

[16] Van Der Meersche, Eddy (2014). *Kristallformen von Fluorit- Crystal forms of fluorite*. Eddy Van Der Meersche. p. 1-296. ISBN 978-9-074669-00-9.

[17] Morgan, Bob. "The fluorite twin morphologies". *The 44th Rochester Mineralogical Symposium, 28.*

[18] Trinkler, M., Monecke, T. and Thomas, R. (2005). "Constraints on the genesis of yellow fluorite in hydrothermal barite-fluorite veins of the erzgebirge, eastern germany: evidence from optical absorption spectroscopy, rare-earth-element data, and fluid-inclusion investigations.". *The Canadian Mineralogist, 43, 883-898.*

[19] Clarke, Edward Daniel (1819). "Account of a newly discovered variety of green fluor spar, of very uncommon beauty, and with remarkable properties of colour and phosphorescence". *The Annals of Philosophy, 14, 34-36.*

[20] Escribano Bombin, M., Lopez Jimeno, C. and Mataix Gonzalez, C. (2019). *Handbook of critical and strategic minerals in the new economy*. Engineering Projects Group. ETSIM, Madrid. p. 253-258.

[21] Gonzalez-Partida, E., Camprub^ A., Carrillo-Chavez, A., D^az-Carreno, E.H.; Gonzalez-Ruiz, L.E.; Farfan-Panama, J.L., Cienfuegos-Alvarado, E., Morales-Puente, P. and Vazquez-Ramnrez, J.T. (2019). "Giant Fluorite Mineralization in Central Mexico by Means of Exceptionally Low Salinity Fluids: An Unusual Style among MVT Deposits". *Minerals, 9, 35*. doi:10.3390/min9010035.

[22] Rebollar, Miguel (2006). *Minerals and Mines of Spain. Vol. III. Halides*. Museo de Ciencias Naturales de Alava. p. 158-224. ISBN 84-7821-633-2.

[23] Ford, Trevor D. (1994). "Blue John fluospar. *Geology Today, 10, (5), 186-190.*

[24] Calvo, Guiomar, and Calvo, Miguel (2006). "Fluorite from Spain. Every colour under the sun". In Jesse Fisher, ed. *Fluorite. The collector's choice*. Lithographie, LLC. p. 38-42.

[25] Garda Garda, Gonzalo and Calvo Rebollar, Miguel (1998).

"Mineralogy of the Asturian fluorite deposits". *Bocamina (4) 50-86.*

[26] Calvo Rebollar, Miguel. *Minerals and Mines of Spain. Vol. III. Halogenides.* Museum of Natural Sciences of Alava. Vitoria. p. 173-176. ISBN 84-7821-633-2.

[27] Calvo Rebollar, Miguel (2008). *Minerales de Aragon.* Prames, Zaragoza. p. 115117. ISBN 978-84-8321-253-0.

[28]Gautron, Laurent (1999). "La fluorite dans le massif du Mont Blanc". *Le Regne Mineral, hors série V, 53-56.*

[29] Bill, H., Sierro, J. and Lacroix, R. (1967). "Origin of the coloration in some fluorites". *The American Mineralogist, 52, 1003-1008.*

[30] Morin, Herve (2010). "Une fluorite du Mont-Blanc classee "tresor national"". *Le Monde, 9 April 2010.*

[31] Belsher, D.O. (1982). "Pink octahedral fluorite from Peru". *The Mineralogical Record, 13, 29-30 + 38.*

[32] Jurgeit, M. and Megaw, P. (2006). "Mexico: Hundred of localities , great associations". In Jesse Fisher, ed. *Fluorite. The Collector's Choice.* Litographie LLC. p. 72-77.

[33] Galindo, C., Baldo, E.G., Pankhurst, R.J., Casquet, C., Rapela, C.W. and Saavedra, J. (1996). "Age and origin of fluorite from the La Nueva deposit (Cabalango, Cordoba, Argentina) based on radiogenic isotope geochemistry (Nd and Sm)". *Geogaceta, 19, 67-69.*

[34] Lira, Raul, and Colombo, Fernando (2014). "Las Especies Minerales". *Relatorio delXIX Congreso Geologico Argentino.* p. 1079-1159.

MOLYBDENITE

[35] Sutulov, Alexander. Molybdenum. Santiago, Chile. Editorial Universitaria, 1962.
214 p.

[36] International Molybdenum Association. Moly Information Center. [en Hnea] < http://www.moly.imoa.info/ > [accessed: 3 January 2007] [in Hnea] < http://www.moly.imoa.info/ > [accessed: 3 January 2007].

[37] Prada, Ricardo and Andreu, Paulino. Catalysts and Adsorbents for the Protection of the Environment: Combustibles Fosiles. [slides] Caracas, Venezuela. 1998. Microsoft PowerPoint presentation.

ACTIVATED CARBON

[38] Moreno C. C, Carrasco M. F., Lopez R., M., Alvarez M. A., "Chemical physical activation of olive mil waste water to produce activated carbons", (2001), Carbon 39 1415-1420.

[39] Moreno C. C., Ferro G. M. A., Rivera U. J., Joly, J. P. C., (1994), Carbon 32, 93.

[40] Reinoso F. Harry M. "Activated Carbon", (2006), Elsevier science & technology books

[41] Reinoso, F., "Carbon activado: estructura, preparation y aplicaciones", (2005), Revista uniandes, Colombia, 66-69.

[42] Radovic L.R., Moreno C. C., Rivera U. J., "Carbon materials as adsobents inaAqueousnm", (2001), Solutions in chemistry and physics of carbon, vol. 27 pp. 227405.

[43] Luna D., Gonzalez A., Gordon M., "Obtaining activated carbon from coconut husk" (2007).

[44] Figueiredo, "V curso", (2004), Taller Iberoamericano

[45] Boehm, H. "Some aspects of the surface chemistry of carbon blanck and other carbons", (1994), Carbon 32,759 - 769.

[46] Dubinin M. M., Zaverina E.D., Radushkevich, L.V., "Sorption and structure of active carbons. I. adsorption oforganic vapors", (2005), J. Phys. Chem 21, 1351-1362.

[47] Higgs, T. W., "Technical guide for environment management of cyanide in mining", (1992), British columbia technical and research committee on reclamation cyanide subcommittee.

[48] Young C. A., Cashin S. P. y Jordan, T. S., "Remediation technologies forthe separation and destruction of aqueous cyanide species", (1996), Prepint 96-149, SME.

[49] Ingles J y Scott J. S., "State of art of processes for treatment of gold mill effluents",

(1987) Mining, mineral and metallurgical processes division, Industrial programs brandch, Ottawwa, Canada.

[50] Devuyst, E. A., Conrad, B. R., "A cyanide removal process using sulfur dioxide and air", (1989), Overview, JOM, December, volume 41, Issue 12, pp. 43 -45.

[51] Smith, A. y Mudder, T., "The chemistry and treatment of cyanidation wastes". (1991),
Mining journal books limited, chapters 1, 2, 5 and 6. London, England.

[52] Whitlok, J. , "The advantages of biodegradation of cyanides', (1989), JOM, No.
December, 46-47.

[53] Botz, M. M. y Stevenson, J. A., "Cyanide, recovery and destruction", (1995), Engineering and metallurgical journal, June, 44 - 53.

[54] Gurol M. C. y Bremen, W. M. y Holden T. E., "Oxidation of cyanides in industrial wastewater's by ozone", (1984), Conference on cyanide and environment proceedings, Tucson, Arizona, 341-361.

[55] Reffas, A.; Bernardet, V.; David, B.; Reinert, L.; Lehocine, B.; Dubois, M.; Btisse, N.; Duclaux, L. 2010. Carbons prepared from coffee grounds by H3PO4 activation: Characteriza-tion and adsorption of methylene blue and Nylosan Red N-2RBL. Journal of Hazardous Materials 175(1-3): 779-788.

[56] Ma, X.; Ouyang, F. 2013. Adsorption properties of biomass-based activated carbon prepared with spent coffee grounds and pomelo skin by phosphoric acid activation. Applied Surface Science 268: 566-570.

[57] Rattanapan, S.; Srikram, J.; Kongsune, P. 2017. Adsorption of methyl orange on coffee grounds activated carbon. Energy Procedia 138: 949-954.

[58] Shamsuddin, M.; Yusoff, N.; Sulaiman, M. 2016. Synthesis and characterization of activated carbon produced from kenaf core fiber using H3PO4 activation. Procedia Chemistry 19: 558-565.

[59] Imessaoudene, D.; Hanini, S.; Bouzidi, A.; Ararem, A. 2016. Kinetic and thermodynamic study of cobalt adsorption by spent coffee. Desalination and Water Treatment 57(13): 6116-6123.

Printed by Books on Demand GmbH, Norderstedt / Germany